Ralph Buchsbaum was born on 2 January
1907 in Oklahoma, the son of a doctor who
was very interested in zoology. Educated at
the University of Chicago, where he obtained his Ph.D. in zoology, he held the
positions of Assistant, Instructor, and Assistant Professor of Zoology. During the
war he was a Captain in the U.S. Air Force
and assigned to the Arctic, Desert, and
Tropic Information Centre. From 1946 to
1950 he was Research Associate in Zoology
in the Institute of Radiobiology and Bio-
physics at the University of Chicago. He is
at present Professor of Zoology, University
of Pittsburgh, a member of many American
scientific associations and also a member
of the Marine Biological Association of the
United Kingdom. He is married and has
two children.

ANIMALS WITHOUT BACKBONES

AN INTRODUCTION TO THE INVERTEBRATES

RALPH BUCHSBAUM

**VOLUME
ONE**

PENGUIN BOOKS

Penguin Books Ltd, Harmondsworth, Middlesex, England
Penguin Books Australia Ltd, Ringwood, Victoria, Australia

—

First published in U.S.A. 1938
Revised edition 1948
Published in Pelican Books 1951
Reprinted 1953, 1955, 1957, 1959, 1961, 1963, 1964, 1966, 1968, 1971, 1972

—

—

Made and printed in Great Britain
by C. Nicholls & Company Ltd
Set in Monotype Times

CONTENTS TO VOLUME ONE

*A complete Index to both volumes
will be found at the end of Volume 2*

TO W. C. ALLEE

WHO INTRODUCED ME TO THE
INVERTEBRATES

PREFACE

TO THE ORIGINAL (AMERICAN) EDITION

ELEMENTARY and general accounts of the invertebrates, suitable for the beginning college student or layman, have been limited to two sorts of books: natural histories, which describe the habits of a great many animals but are lacking in descriptions of basic structure and in theory, and formal textbooks, which are packed with morphological detail and technical terminology. This book is an attempt to present the main groups of invertebrate animals in simple non-technical language. Each group is used to illustrate some principle of biology or some level in the evolution of animals from simple to complex forms. The material is divided into two kinds of chapters. The basic or *indispensable chapters* (1, 2, 3, 4, 6, 7, 10, 11, 14, 17, *19, 22, 25, and 28) present a continuous story of the more general and elementary aspects of the main invertebrate groups and can be read consecutively without reference to the others. They are recommended for the general reader or for students in any introductory college course in biology. They are essentially the ones read by our students for the invertebrate section of the 'Introductory General Course in Biology' at the University of Chicago. The remaining chapters are advanced or *optional chapters*, and these treat each group more fully and present additional principles. They are intended to serve as optional reading for the invertebrate section of a general course. The book as a whole is designed as a textbook for a college course in invertebrate zoology.

In the preparation of the manuscript I was extremely fortunate in having the advice and criticism of Dr Libbie Hyman, whose extensive knowledge of the literature on the lower invertebrates has helped to bring the material up to date. She has contributed many valuable suggestions to the organization of the book and has corrected many errors. To Dr Merle Coulter, of the Department of Botany and director of the Introductory General Course in Biology at the University of Chicago, I am indebted for the idea of dividing the book into two sets of chapters. Dr Coulter

* Chapters 19, 22, 25, and 28, *see* Volume Two.

has read the manuscript; and his suggestions, free from the natural bias with which a zoologist approaches a book on animals, have been especially helpful in improving the clarity of many sections. Dr W. C. Allee, of the Department of Zoology, has read the manuscript critically, much to its advantage, and has supplied encouragement throughout the preparation of the book. From Dr Alfred Emerson, of the Department of Zoology, I have adopted many ideas for the organization of the material. Dr Emerson has read the manuscript and has made many suggestions, particularly in the chapters on the arthropods. Dr Thomas Park, of the Department of Zoology, contributed a valuable criticism of the chapters on the arthropods. Dr Harry Andrews, of the Chicago Junior Colleges, has read some of the early chapters. To all of these colleagues and friends, who have given so generously of their time and effort, I am deeply indebted and sincerely grateful.

The drawings are mostly diagrammatic and have been designed to convey ideas about structure, function, or habit, rather than to show the details of any particular species of animal (except where specifically labelled). As far as possible, the same symbols have been used for corresponding structures in different drawings; and it is hoped that this will aid in their ready interpretation. The large number of new drawings, and the adaptation of the borrowed ones to the style followed in this book, required a close collaboration between author and artist made possible only by the fortunate circumstance that the artist was my sister, Miss Elizabeth Buchsbaum. To her skilful and artistic execution of the drawings the book owes much of its attractiveness.

The numerous photographs, unusual in a textbook, supply the elements of specific form and texture which are missing from the diagrammatic, stylized drawings. They are intended as a sort of laboratory exhibit and vicarious field experience.

The source of every photograph is acknowledged in the legend which accompanies it, but I am especially indebted to Dr Douglas Wilson, of the Plymouth Laboratory of the Marine Biological Association, England, for several excellent pictures from his book, *Life of the Shore and Shallow Sea*; Dr W. K. Fisher, of the Hopkins Marine Station, Pacific Grove, California; Mr Richard Westwood, managing editor of *Nature Magazine*; Mr A. S. Windsor, of the General Biological Supply House, Inc., Chicago; Dr C. M. Yonge, for photographs from his book, *A Year on the Great Barrier Reef*; Dr James A. Miller, of the Department of Anatomy, University of Michigan; Mr Albert Galigher, of Berkeley, California; Mr Louis Diamond, Chicago; and Mr Leon Keinigsberg, Chicago.

Mr Percival S. Tice, of Chicago, deserves special mention for his outstanding photographs, many of which he made especially for this book. I value them particularly because some of them, like the hydra series, are of subjects extremely difficult to photograph and not usually attempted.

All photographs not otherwise acknowledged are by myself, and I take this opportunity to express my thanks for facilities and assistance offered me at the various marine biological stations where I have made photographs. I am especially indebted to Dr W. K. Fisher, director of the Hopkins Marine Station, Pacific Grove, California, and Dr Bolin of the same station; Dr J. F. G. Wheeler, director of the Bermuda Biological Station, St George's, Bermuda; Dr A. Tyler of the Kerckhoff Marine Laboratory, Corona del Mar, California; and officials of the Marine Biological Laboratory, Woods Hole, Mass.

I also wish to thank those who have helped in getting the book through the press. Miss Sylvia Shaffer aided in typing the manuscript and in preparing the Index. Finally, I am indebted most of all to my wife, who has assisted in every phase of the preparation of the book, from writing the text and reading proof to designing drawings and making photographs.

RALPH BUCHSBAUM

University of Chicago
August 1938

NOTE

DR BUCHSBAUM'S *Animals without Backbones* introduces the English reader to a new kind of book – a serious scientific work which is as attractive and easy to read as any natural history. It differs from most textbooks in its rejection of all detail which tends to distract interest and attention from main issues, and in the references it makes to animals that matter to human beings. It is also outstanding pictorially. There are photographs of animals alive and at home, as beautiful technically as the illustrated weeklies have taught us to demand. There are drawings so sure and vivid that you feel the animal moving in its characteristic way. And there are diagrams so bold and graceful that hard facts about the structure of animals are easily taken in. Thus aquarium and laboratory are brought together, and you can watch a starfish living its starfish life on one page, and can study its anatomy with dissecting instruments and microscope on another.

Animals without Backbones is published in two volumes in the Pelican series. Each volume gives a complete account of the various groups of animals it deals with, and so can be read by itself. The two together give a comprehensive survey of the invertebrate members of the animal kingdom, and if you want to take pleasure in discovering the serious scientific content of invertebrate zoology, these are your books.

M. L. JOHNSON

For technical reasons it has been impossible to make the division of the photographic plates between the two volumes of *Animals without Backbones* agree with the division of the text into two volumes. The animals which are illustrated on plates 54–64 in this volume are discussed in the early chapters of volume 2.

CHAPTER 1

By Way of Introduction

ANYONE can tell the difference between a tree and a cow. The tree stands still and shows no signs of perceiving your presence or your hand upon its trunk. The cow moves about and appears to notice your approach. This striking difference in the behaviour of plants and animals is related to the fundamental difference in plant and animal nutrition.

Plants make their own food from simple constituents of the air and soil. With the aid of a green pigment, chlorophyll, the tree utilizes energy from the sun to combine carbon dioxide and water into food – a process known as photosynthesis. The cow cannot stand in the sun and soak up energy with which to make nutritive substances, but must get its food by eating the bodies of the food-manufacturers, the independent plants. To find a constant supply of plants the cow must move from place to place and must react rapidly to changing conditions in its environment.

Not all animals move about. The sponges, for example, grow attached to the substratum. They have internal moving parts which create currents in the water, thus drawing food towards the sponge. Since this was not apparent to the early naturalists, they classified sponges and many other stationary animals as plants. While there is no longer any doubt that the sponge in the bathroom or the piece of coral decorating the mantlepiece are the skeletons of animals, the question 'What is an animal?' is not always easy to answer exactly.

As we examine simpler and simpler forms of life, distinctions of behaviour and of nutrition grow less and less obvious. Eventually we find microscopic organisms that exhibit some characteristics possessed by both plants and animals. These 'plant-animals', the FLAGELLATES, swim about by lashing long thread-like extensions called FLAGELLA (singular, flagellum – the Latin word for 'whip'). Some flagellates carry on photosynthesis, but they move about and show the same sensitivity and rapidity of response as do typical animals. Some flagellates not only photosynthesize but also feed like animals, thereby seeming to make doubly sure of a source of nourishment and aligning themselves with neither plants nor animals. Other flagellates have lost their chlorophyll and feed only in an animal manner. The existence of such organisms indicates that, in the beginning, there were no differences between plants and animals and that life was restricted to very simple forms. What these forms were and how they originated are questions on which we have no direct evidence.

One of the most plausible of the hypotheses advanced to account for the origin of life states that at some time in the earth's history, in suitable places, as in ponds or on the seacoasts, there were, as there are now, simple compounds of the elements which compose the living substance, protoplasm. With the energy of the sun or the heat of warm springs, various chemical combinations were formed. Some of these possessed the power of self-propagation, that is, the ability to manufacture additional combinations like themselves. An analogy to such a state of living matter may be found in a group of substances which are so small that they are just under the limits of visibility of the ordinary microscope and pass through the pores of the finest porcelain filters. Because of this filter-passing character, and because they are responsible for diseases such as smallpox, yellow fever, mumps, infantile paralysis, and the mosaic diseases of plants, these substances are called *filterable viruses* ('poisons'). The viruses are among the largest proteins known, and several different ones have already been prepared in pure crystalline form. Even after repeated crystallizations, a treatment no obviously living substance has ever been able to survive, viruses resume their activities and multiply when returned to favourable conditions. While no one has yet succeeded in growing them in the absence of living matter, it is clear that viruses help to bridge the gap that was formerly thought to exist between nonliving and living things. No longer can it be said that there is some sharp and mysterious distinction between the nonliving and the living, but rather there seems to be a gradual transition in complexity.

If we imagine that the earliest self-propagating substances were something like viruses, it is not difficult to suppose that an aggregation of virus-like proteins could lead to the development of larger bacteria-like organisms, independent, creating their own food from simple substances, and using energy from the sun.

Such a level of organization may be compared to present-day forms like the *independent bacteria*, some of which conduct photosynthesis without chlorophyll, using, instead, various green or purple pigments. Others utilize the energy derived from the oxidation of simple salts of nitrogen, sulphur, or iron. These, for instance, can oxidize ammonia to nitrates, or hydrogen sulphide to sulphates, with the release of energy which is utilized in forming carbohydrates.

From primitive bacteria-like forms to the simple chlorophyll-bearing organisms is a relatively short jump in complexity, however long it may have taken in time.

There is evidence that both plant and animal kingdoms originated from primitive flagellates. By losing locomotor flagella and assuming a rounded form, some flagellates become indistinguishable from the simplest plants, the algae; and, in fact, many of the green flagellates regularly pass into such an immotile state when they reproduce. By loss of chlorophyll, other flagellates become purely animal types which capture and ingest food.

Arising from primitive flagellates, animals have evolved into a bewildering variety of forms of ever increasing complexity of structure. When these animals are carefully studied and compared, it is found that many of them resemble man in various ways, notably in the presence of a row of bones (vertebras) along the middle of the back, as well as in the presence of bones inside the limbs and head. The animals having internal bones, including a backbone or vertebral column, are known as VERTEBRATES, and comprise all the fish, the frogs, toads, and salamanders, the lizards, snakes, turtles and crocodiles, every kind of bird and all the hairy animals known as mammals, such as elephants, lions, dogs, bats, and mice. These more or less familiar animals have a highly exaggerated importance in our minds because they are closely related to man, because they are mostly of large size, and because, like man, they usually manage to make themselves conspicuous. Actually, in terms of number of species they comprise only about 5 per cent of the animal kingdom.

The remaining 95 per cent consists of animals without back-bones. We are all aware of the difference between these two groups of animals when we indulge in fish and lobster dinners. In the fish the exterior is relatively soft and inviting, but the interior presents numerous hard bones. In the lobster, on the contrary, the exterior consists of a formidable hard covering, but within this initial handicap is a soft edible interior. A similar situation exists in the oyster, lying soft and defenceless within its hard outer shell. The lobster and the oyster are but samples of a tremendous array of animals which lack internal bones and which are, from lack of the vertebral column in particular, called INVERTEBRATES.

A distinction between vertebrates and invertebrates was first recognized by Aristotle, although he did not use these terms but divided animals into those with blood (vertebrates) and those without blood (invertebrates). Unfortunately, Aristotle's neat distinction had little to do with the facts, since many invertebrates possess red blood and a great many other invertebrates have colourless blood, which he did not recognize as blood at all. Although Aristotle did about as well as one might expect from the limited knowledge of animal structure available in his time, it was partly because of the weight of his authority that his error was not corrected for over two thousand years. With the development of the fruitful independent scientific spirit of the early nineteenth century, Lamarck and Cuvier finally made the correct distinction, based upon the fundamental plan of organization of the animal body.

There is a popular but vague recognition of the difference between vertebrate and invertebrate animals in the expression 'spineless as a jellyfish'. In this book we shall be concerned not only with the jellyfish, which is seldom seen by inland dwellers, but also with many animals without backbones, like clams, crayfish, earthworms, and fleas, which are supposed to be already familiar to most people. In addition, many forms will be presented which generally pass unnoticed because they are too small to be seen without a microscope, because they live under water or in the ground, because they inhabit remote parts of the world, or simply because they escape the unobservant eye.

CHAPTER 2

Life-activities

To keep alive and healthy, all animals from the lowest to the highest must carry on certain LIFE-ACTIVITIES. Because these activities centre about the utilization of energy, they have often, and very appropriately, been compared to the functioning of a combustion engine. But there is a point at which this analogy breaks down. For not only is the living machine a self-feeding, self-tending, and self-perpetuating one, but by its very nature is a machine which must operate at all times. A stalled motor may easily be repaired. Not so a 'stalled' organism, in which the failure of certain functions automatically brings on the disintegration of the machinery itself. A machine can be oiled, covered, and put away on a shelf until ready for use, while an organism must be kept running, sometimes rapidly, sometimes very slowly, but continuously from the start until the natural or accidental finish.

This continuity of living processes appears, on first thought, to be contradicted by our ordinary observations of animals. For example, we know that crayfish may lie dormant in the mud of a dried-up pond and then resume their activities when spring rains fill the pond. What we see here is only a temporary cessation in the easily observable activities of the crayfish. Going on all the time at a greatly slowed rate is a much more fundamental activity, the liberation of energy. The energy liberated comes from the burning of food stored by the crayfish during the months preceding dormancy. This food was obtained directly from plants or 'second hand' from other animals that had eaten plants. To catch, bite off, and chew up parts of other animals or of plants, the crayfish has special structures and behaviour which make possible an activity lacking in plants and so characteristic of animals – the capture and ingestion of food.

INGESTION is the taking-in of food. Animals differ strikingly in their mode of ingestion. The differences are related partly to the size and complexity of the animal and partly to the diversity of the food itself. Mouth parts that tear flesh will not do for chewing wood; sucking sap is not the same as sucking blood. In its essentials, however, the feeding machinery involves: a set of *sensory receptors* to get information about the external environment, *mechanisms for locomotion and ingestion*, and *a means of co-ordinating* the locomotory and feeding mechanisms with the information received from the surroundings so that the net result will be the getting of something to eat and the avoidance of being eaten.

DIGESTION is the chemical alteration of raw food into a form in which it is usable as a source of *energy* for the life-activities and as a source of *materials* for growth and replacement of worn-out or damaged parts. The raw food consists of water, carbohydrates, fats, proteins, inorganic salts, and some other substances. The water and some of the salts need not be digested; they are immediately available for incorporation into the living animal body. The other substances must be broken down into simpler units because they are too large to pass through the living membranes and because they are too complex to be used directly in growth or in other living processes. Digestion, then, is the breakdown of raw food into smaller units. To facilitate this breakdown, the living organism has a digestive apparatus into which are poured a number of kinds of chemical substances, most important among which are the ENZYMES.

Enzymes are complex substances, manufactured only by living organisms, which speed up chemical reactions. Enzymes are notably specific, that is, they accelerate only one particular reaction. For example, some digestive enzymes act only on carbohydrates, others only on fats, and still others only on proteins.

ELIMINATION is the ejection from the body of indigestible food or other accumulated solid wastes. Most plant-eaters do not have the enzymes needed to digest completely the woody tissues of the plants they feed upon. Most insect-feeders cannot break down the complex substance that forms the hard outer skeleton of insects. These indigestible portions of the food constitute the solid wastes or *faeces*, and must be removed lest they clutter up the digestive machinery

METABOLISM is the total of the chemical changes that go on in the animal body. There are two phases of metabolism: building up, or *constructive metabolism;* and breaking down, or *destructive metabolism*. During the period of active growth and repair the constructive phase overbalances the destructive phase, whereas during old age or in disease the destructive phase predominates. One kind of history of an animal may be told in terms of the metabolic changes or events that befall it from beginning to end.

ASSIMILATION is a constructive metabolic process by which materials derived from digestion are incorporated into the living substance, protoplasm. After the food is converted from its condition as the structural part of one kind of animal or plant into smaller, simpler units, it is suitable for building the kinds of carbohydrates, fats, and proteins peculiar to the structure of the animal concerned. Just as innumerable useful objects can be fashioned from combinations in various proportions of only a few dozen building-materials, so a few dozen kinds of food units can be built into an almost infinite variety of organisms – each kind of organism with its own specific composition.

RESPIRATION is a destructive chemical process by which food is burned in the release of energy. The energy stored in the food through the photosynthetic action of green plants is released in somewhat the same way that man releases, by burning, the energy stored in coal. The high temperatures involved in the burning of coal are not necessary in respiration because the chemical reactions are accelerated not by heat but by special respiratory enzymes. The burning of coal, or of almost anything else, requires air, or, more exactly, the oxygen of the air. The release of energy in the living organism may be described in three steps. The first is the bringing of oxygen to the fuel. In man this is accomplished by *breathing* and by the circulation of the blood from the lungs to the tissues. The second step is the actual burning, or the chemical union of oxygen with the fuel, resulting in the liberation of energy; this is known as *oxidation*. The third step is the removal of the by-products of respiration – water and carbon dioxide. These wastes usually pass in the reverse direction along the same route by which the oxygen entered.

EXCRETION is the separation from the living protoplasm of the waste products of assimilation and other chemical activities. The by-products of oxidation of carbohydrates and fats are carbon dioxide and water. The burning of proteins yields other wastes in addition to carbon dioxide and water, namely, compounds of nitrogen, which require special methods of disposal. These nitrogenous compounds are poisonous, and their prompt removal is indispensable to life. In man and in other vertebrates the kidneys filter out these wastes in the formation of urine, which is eliminated to the outside; various devices do this same work for other animals.

REPRODUCTION is the production of new individuals to take the places of the old ones which die because their machinery wears out or because they are eaten or destroyed by their enemies. Unlike the other life-activities, reproduction is not necessary to maintain life in any single individual; it is essential only for the continued existence of the group.

WHILE the mechanisms employed by animals for carrying on their life-activities differ considerably, they are all made from variations of a basic living substance called PROTOPLASM.

The protoplasm of larger animals does not exist as a continuous mass but is divided up by partitions into minute units called CELLS. When viewed through the microscope, each cell appears as a bit of protoplasm enclosed in a CELL MEMBRANE which separates it from adjacent cells. In nearly all plants the cells are surrounded by heavy walls made of a carbohydrate, cellulose. With very few exceptions, animal cells lack heavy walls, the cell membrane being a thin, flexible layer composed largely of fat and

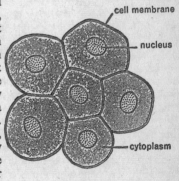

cell membrane

nucleus

cytoplasm

A group of CELLS

protein. The difference in the chemical structure of plant and animal cell membranes is correlated with motility. The protoplasm of the cell is differentiated into a centrally located body, the NUCLEUS, and the remaining protoplasm, called CYTOPLASM, surrounding the nucleus. The cell is the unit of living structure and activity. This is the same as saying that the cell is the smallest part of an animal which can carry on all of the life-activities.

If we remove a small piece of tissue from an animal, as from a man, cut it into small fragments, and keep these fragments in a suitable fluid, we find that parts of the cells disintegrate, but that intact cells will go on metabolizing and reproducing so long as they are properly cared for. However, without this care (a technique known as *tissue culture*) these cells die, because they are so specialized to perform particular activities that, when they are removed from the organism in which they are a part of a co-ordinated whole, they cannot take up an independent existence in the external world.

There are free-living organisms whose bodies are not divided up into cells. Instead, they consist of a continuous mass of protoplasm enclosed in a single membrane and containing one or more nuclei. Whatever differentiation is present has occurred within the protoplasm of a single microscopic cell.

BEGINNING with one of the least complex of these unicellular animals, we shall see, in the chapters that follow, the ever-increasing complexity and efficiency of the living machinery with which the various kinds of invertebrates carry on their life-activities. The details of animal structure which will be presented may be interesting in themselves, but are meaningless unless we view them in relation to their function in the life of the animal and as a stage in the evolution from simple to complex forms.

CHAPTER 3

The First True Animals

THE microscopic unicellular animals constitute the first of the large groups of PHYLA into which the animal kingdom is divided. This group is called the phylum PROTOZOA, a name which means 'first animals'. Of the thousands of species of protozoa that occur everywhere in fresh and salt waters, damp soils, and dry sand, and that live upon or inside the bodies of other animals as parasites, only two will be discussed in this chapter – one simple and one more complex form.

The relation between a simple animal and a complex animal may be compared to the relation between the behaviour of prehistoric man and the behaviour of modern man. The first man, we may imagine, lived without mechanical aids of any kind, captured his food with his bare hands, sought shelter in a tree or a cave, and existed at the mercy of natural forces. Modern man, on the other hand, has invented mechanical devices to assist him in every way, and is gradually learning to protect himself from the caprices of nature. In an analogous way, a simple animal represents a level of organization in which the protoplasm has not evolved many special devices making for greater efficiency in the business of life. Such an animal lives on what we may call the PROTOPLASMIC LEVEL OF CONSTRUCTION; the protoplasm performs all the life-activities and does not have any very complex structure correlated with any particular activity.

The appearance of specific structures which add to the efficiency of performance of some special activity is called *differentiation* or *specialization*. The amoeba is an animal that displays a minimum of differentiation.

A SIMPLE PROTOZOAN – AMOEBA

THE common amoeba of fresh-water ponds is a microscopic animal; but some of the largest known amoebas may reach a diameter of half a millimetre, being visible to the naked eye as

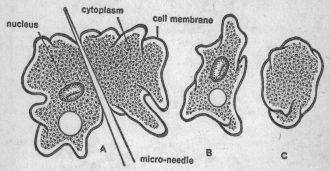

A, an AMOEBA IS CUT IN TWO with a very fine glass needle. B, the piece containing the nucleus. C, the piece without the nucleus.

white specks. Each amoeba is a little mass of clear gelatinous protoplasm containing many granules and droplets. The surface of the amoeba's protoplasm forms a delicate CELL MEMBRANE through which materials pass in and out of the animal. Water passes freely through the cell membrane; but the proteins, carbohydrates, fats, and salts of the protoplasm are prevented from escaping into the surrounding water by this same membrane. If an amoeba is cut in two, each piece rounds up and immediately produces a complete membrane, thereby preventing the loss of the interior protoplasm. The formation of a surface membrane appears to be a general property of protoplasm.

The protoplasm of the amoeba, as in almost all cells, is differentiated into nucleus and cytoplasm. The NUCLEUS occupies no fixed position. The amoeba furnishes excellent material for the

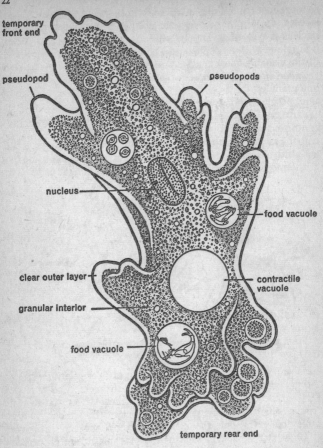

An AMOEBA, showing principal structures.

study of the function of the nucleus, since the animal may be cut into two pieces, one with and the other without the nucleus. The piece with the nucleus behaves like an entire animal, soon grows to its previous size, and finally reproduces. The piece without the nucleus moves about in more or less normal fashion for a time and may sometimes feed, but it is unable to digest food, to grow or to reproduce; and it dies after the food stored in the protoplasm is used up. From such experiments it is concluded that the nucleus is concerned largely with constructive phases of metabolism and with reproduction. The CYTOPLASM of the amoeba is distinguishable into a *clear, outer layer* and a more or less *granular interior*. This interior contains various sorts of crystalline granules, fat droplets, and food bodies in process of digestion, besides droplets containing a watery fluid.

MOTILITY is one of the striking characteristics of animals as contrasted with plants. The type of movement exhibited by the amoeba is called, naturally enough, 'amoeboid movement', and has always excited great interest because it is presumed to be one of the most primitive types of animal locomotion. It appears to

An AMOEBA IN PROFILE. When the microscope is arranged so that the animal is viewed from the side, it can be seen that only the tips of the pseudopods are in contact with objects, the general mass being free in the water. The pseudopods appear to act like little legs put out one after another, but the 'legs' are temporary and soon flow back into the general cytoplasm. (Based on Dellinger.)

be totally different from the muscular movements of complex animals; but what goes on in a muscle when it contracts to move a limb may prove to be similar to the chemical and physical changes that go on in a moving amoeba. Furthermore, some of the cells in the tissues of all higher animals, including man, are amoeboid.

For these reasons AMOEBOID MOVEMENT has been the object of intensive study. Amoebas have no distinct head or tail ends but have a surface which is everywhere the same, and any one point on this surface may flow out as a blunt projection or PSEUDOPOD ('false foot'). This pseudopod continues to advance for some time through the passage into it of some of the mass of the amoeba, but sooner or later another similar projection forms at an adjacent point, and then the cytoplasm flows into the new pseudopod. In this manner the animal progresses in an irregular fashion – flowing first to one side, then to the other. It often alters its course by putting out pseudopods on the side opposite the previous advance. As new pseudopods form, the old ones flow back into the general mass. Amoeboid movement is very slow, and the animal does not proceed for long in any one direction.

Of the various explanations of amoeboid movement that have been advanced, the one which seems most acceptable at present is based on changes in the consistency of the cytoplasm. The cytoplasm, as already noted, is visibly differentiated into a clear, outer layer and a granular, inner one. In the inner layer we further distinguish an outer jellylike region, the PLASMAGEL, and an inner fluid region, the PLASMASOL. In a moving amoeba plasmasol flows in the direction of movement; as it reaches the tip of the pseudopod and is deflected to the sides, it changes to plasmagel, while more plasmasol flows forward into the moving tip. This process can be compared to a stream of liquid cement in which the

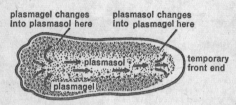

The MOVEMENT OF AN AMOEBA depends upon reversible changes in the consistency of the cytoplasm. The arrows indicate the path of flowing cytoplasm. (Modified after Mast.)

flowing cement hardens on the outside, building ahead of itself a tube of hard cement through which comes more liquid cement from behind. A constant flow of plasmasol is maintained because the plasmagel liquefies at the rear and again flows forward as plasmasol. The flow occurs probably because the tube of plasmagel contracts, exerting pressure on the fluid interior and forcing it forward at the point where the plasmagel is thinnest. Any influence which causes the plasmagel to thicken at the forward end will change the direction of movement. Thus, if pressure is exerted against the end of the leading pseudopod, a thick layer of plasmagel forms at this end and the flow of the animal is reversed. An amoeba that is irritated simultaneously on several sides rounds up into a motionless ball, presumably because a heavy layer of plasmagel forms on all sides.

Pseudopod formation occurs not only in locomotion but also in the capture and INGESTION of food. Although they have no special cell organs of taste or smell, amoebas are able to distinguish inert particles from the minute plants and animals upon which they feed. Pseudopods are thrown out around the sides and over the top of the object. In this way the food is held against the substratum, then completely surrounded by cytoplasm, and finally incorporated into the main mass of the amoeba's body. The behaviour of the amoeba varies somewhat with the kind of food. If the food organism is active, the pseudopods are thrown out widely and do not touch or irritate the prey before it has been surrounded. When the amoeba is ingesting a quiescent object, such as a single alga cell, the pseudopods surround the cell very closely.

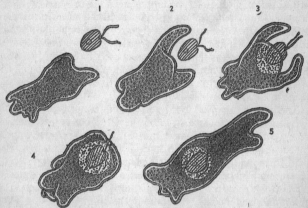

AN AMOEBA INGESTS a flagellate. 1, the amoeba moves towards the prey. 2, pseudopods begin to extend. 3, pseudopods are thrown out around the sides, and a thin sheet of cytoplasm extends over the top. 4, a sheet of cytoplasm extends below the prey, which is now completely enclosed except for one flagellum. 5, the food organism lies in the amoeba's cytoplasm in a food vacuole. (Based on Schaeffer.)

The food body usually lies in the amoeba's cytoplasm in a drop of water which was taken in when the food was enclosed by the pseudopods. This drop of water containing the food is called a FOOD VACUOLE. The food very soon begins to undergo DIGESTION, presumably by enzymes that enter the food vacuole from the surrounding cytoplasm.

Enzymes act best at a definite acidity or alkalinity; consequently, the amoeba must not only provide enzymes for the digestion of food but must furnish these enzymes with proper conditions. In the food vacuoles of the amoeba the reaction is at first acid, the acidity serving to kill the prey, which often struggles for some time after being enclosed. The reaction later becomes alkaline; for the most important enzyme, the protein-digesting enzyme of the amoeba, is active only in an alkaline medium. This reminds us, in a general way, of the situation in man where food is first acted upon in the acid medium of the stomach and later in the alkaline condition of the small intestine.

As the food body gradually dissolves, the dissolved substances pass into the general cytoplasm, where they are ASSIMILATED. The indigestible fragments are ELIMINATED in the simplest fashion possible. They are gradually shifted to the temporary rear end of the animal and are left behind as the amoeba flows away.

RESPIRATION requires no special breathing mechanism in a minute creature like an amoeba. There is free exchange of gases with the surrounding water, and the amoeba does not 'breathe' in the same sense as this expression is used with regard to man, that is, its sides do not heave in and out. Yet it carries on all the essentials of respiration in that energy is liberated from food and made available for other life-processes.

The oxygen dissolved in the surrounding water passes into the cytoplasm of the amoeba by diffusion. (*Diffusion* is the tendency of the particles, of which matter is composed, to disperse equally throughout any space which is available; that is, the particles tend to move from regions of higher to regions of lower concentration.) Since the concentration of oxygen in the water is higher than that in the amoeba's cytoplasm, oxygen constantly enters and is immediately used up in the burning of foods. Thus the concentration of oxygen within the animal always remains lower than that in the outside water, and oxygen continuously enters the animal and is available for energy requirements.

The water that results from the burning of carbohydrates and fats is a normal constituent of the animal body, and its rapid disposal is not necessary. The carbon dioxide is harmful if allowed to accumulate, and must be removed more promptly. It diffuses to the outside because it is always at a higher concentration within the amoeba than in the surrounding water. This method of respiratory exchange (diffusion in of oxygen and diffusion out of carbon dioxide) will work successfully only if the animal is very minute, so that its exposed surface is large in proportion to its bulk or mass, and if it is not covered with a thick protective layer which would interfere with the free diffusion of gases.

Burning protein, as we already know, yields not only carbon

dioxide and water, but also nitrogenous substances which are poisonous and which must be rapidly EXCRETED. In the amoeba no special parts have been shown to excrete harmful wastes and these are thought to diffuse through the cell membrane into the surrounding medium.

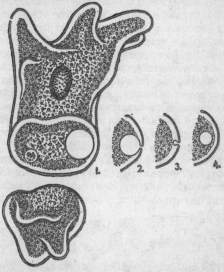

The CONTRACTILE VACUOLE as it would appear if we could make a crosswise cut through the amoeba at the level of the vacuole and move the rear piece back without the contents of the amoeba spilling out. 1, the vacuole at maximum size. 2, ejecting its contents to the outside through a minute pore. 3, almost completely emptied. 4, the vacuole forms again and increases in size. The pore is a temporary structure, formed anew for each ejection.

Near the rear of the moving amoeba is a large spherical water vacuole, called the CONTRACTILE VACUOLE, which contracts at regular intervals, discharging its contents to the exterior. It then forms again from one or more minute droplets and gradually swells to a maximum size, whereupon it again collapses, ejecting its contents through a temporary pore to the outside.

The role generally assigned to the contractile vacuole is like that of a pump in a leaking ship, in which the pump must be kept going all of the time to keep pace with the incoming water. The water pumped out by the vacuole may come from several sources: it may be produced as a result of respiration, it may be included when food particles are engulfed, or it may enter (by osmosis) because the salts and certain other substances in the protoplasm of the amoeba are more concentrated than those in its fresh-water environment. (In a living system *osmosis* is the diffusion of water through a membrane which is permeable to water but not to most dissolved substances. When two solutions of different concentrations are separated by such a semi-permeable membrane, the water diffuses in both directions, but more rapidly from the less concentrated to the more concentrated solution than in the reverse direction, because in the more concentrated solution the greater number of dissolved particles interferes with, and in other ways decreases, the outward diffusion of the water. As a result, water tends to accumulate on the side with a greater concentration of dissolved substances.) Experimentally increasing the concentration of salts outside of the amoeba causes the vacuole to contract less and less frequently and finally to vanish altogether. Conversely, some marine amoebas develop contractile vacuoles when placed in fresh water. Thus, it is probable that the chief function of the contractile vacuole is to *regulate the water content* of the amoeba.

An apparatus of this kind seems admirably suited for expelling nitrogenous wastes, but no one has yet been able to produce convincing experimental evidence that the vacuole serves an important excretory function.

On the other hand, there is evidence against the excretory role of the contractile vacuole. When the fluid in the vacuole is withdrawn by means of a micropipette, chemical analysis fails to show a concentration of urea (nitrogenous waste) high enough to indicate that the vacuole has a special excretory function. Analysis of vacuole fluids of other protozoa shows that, even though small amounts of nitrogenous materials may be dissolved in the water expelled from the vacuole, the amount is not great enough to account for the total amount of waste excreted by these organisms, and most of the materials must diffuse out through the cell membrane. Perhaps improved chemical methods will settle this problem in the future.

Less direct evidence for the water-regulatory role of the vacuole is the fact that many marine and parasitic protozoa, which live in environments having a salt concentration higher than that in their own protoplasm, do not possess vacuoles. Since water moves through a membrane from a region of low salt concentration to one of higher salt concentration it does not accumulate within these animals. In the parasitic protozoa that have vacuoles, most of the water expelled probably enters in the feeding process.

The amoeba REPRODUCES by the simple process of dividing into two amoebas. When an amoeba has fed well for some time, it rounds up into a spherical mass, the nucleus divides, and the cytoplasm constricts until the slender strand that connects the two halves ruptures. The entire process requires less than an hour.

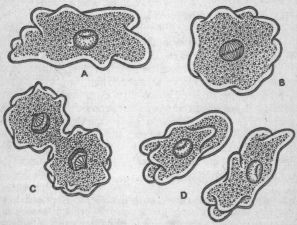

An AMOEBA DIVIDES. A, old well-fed amoeba. B, the amoeba rounds up, and the nucleus undergoes a change preparatory to division. C, the nucleus divides (by a process known as 'mitosis'), and the cytoplasm constricts. D, the two young 'daughter' amoebas, each with a nucleus and half the parent's cytoplasm. (Based on Dawson, Kessler, and Silberstein.)

Each half, called for no particular reason, a 'daughter' amoeba, behaves just like the parent and soon increases to maximum size. Since an amoeba thus continues to exist in its offspring, it may be said to be 'immortal'; and every amoeba which now exists is directly continuous through the ages with the first amoeba. However, the protoplasm of every cell is continually being destroyed and renewed, and we may be almost certain that no part of the original amoeba is present in any modern amoeba.

The amoeba can carry on its routine activities only when immersed in water. If conditions of life become unfavourable, as when the pond dries up or when the food supply runs low, the

An ENCYSTED AMOEBA survives unfavourable conditions. With the return of proper conditions, the animal emerges from its cyst.

amoeba rounds up and secretes on its outer surface a hard and impervious protective shell called a CYST. Within the cyst the animal's rate of metabolism falls to a level just above that necessary to maintain the organization of the protoplasm. Replaced in a suitable environment, the cyst breaks open and the enclosed animal emerges and resumes its usual activities.

Although the amoeba has nothing comparable to our organs of special sense, it can distinguish food from particles of no food value. As mentioned before, it uses different tactics in approaching plant and animal food, probably because the movements of animal prey create disturbances in the water that stimulate the amoeba. The amoeba flows away from a bright light, injurious chemicals, or mechanical injury. When poked with a glass rod, it contracts, reverses the direction of flow, and moves away. Extreme disturbance or injury causes it to take on a spherical shape and remain motionless for some time.

The ACTIVITIES of an amoeba, particularly its ability to select food, have been used by some as evidence that an amoeba exhibits 'conscious' behaviour or possesses some traces of those powers which in man are vested in the brain and which have been called the 'psychic property' of protoplasm. Others maintain that the simple activities of an amoeba imply no psychic attributes, and point to the fact that it is possible to duplicate practically all of the activities of an amoeba with (non-living) MECHANICAL MODELS – not only the chemical changes, such as those involved in respiration and digestion, but also the more characteristically living activities. Amoeboid movements are produced by injecting a little alcohol into a droplet of clove oil in water. The alcohol changes the surface film of the oil droplet and causes it to send out 'pseudopods' and to flow about like an amoeba. A drop of chloroform in water appears to be quite as 'finicky' in its 'eating habits' as an amoeba. When offered small pieces of various substances, such a drop will 'refuse' sand, wood, and glass and will even eject them if they are forcibly introduced. On the other hand, bits of shellac or paraffin are 'eagerly' enveloped. If we play a trick on the chloroform drop by feeding it a piece of glass coated with shellac, it will engulf this 'delicacy', dissolve the shellac, and then 'eliminate' the glass. There are many other mechanical models which simulate growth and even reproduction. The resemblances are usually quite superficial, and most of them throw little light on the real mechanisms involved in the living systems which they apparently imitate. But these experiments do suggest that much of the 'mystery', which some writers attribute to the behaviour of the living amoeba, might be explained if we knew enough about the purely physical phenomena involved.

The amoeba differs from the mechanical models in that several models are required to demonstrate the activities that are displayed by a single amoeba, a fact which emphasizes the complexity of this 'simple animal'. A more important difference is that the behaviour of the amoeba is *adaptive*, that is, it is of a type likely to result in survival of the animal.

In the amoeba we have emphasized simplicity and the ability of protoplasm to perform all the necessary life-activities without the aid of highly specialized structures. We shall see now that in another member of the large and varied phylum Protozoa, many specializations are possible even within the limits of the protoplasm of a single cell. These specializations have the same use as the inventions and machines of human construction: they enable the animal to carry on its activities with greater efficiency.

A COMPLEX PROTOZOAN – PARAMECIUM

PARAMECIA are found everywhere in fresh waters and can be obtained in enormous numbers by letting a bit of food decay in pond water. Like an amoeba, a paramecium consists of a microscopic mass of protoplasm which is differentiated into a semifluid granular interior and a more dense, clear, outer layer. But many differences between the two animals are at once apparent. Instead of a delicate outer membrane, a paramecium is covered by a stiff but flexible OUTER COVERING. This covering gives the animal a definite PERMANENT SHAPE, somewhat like that of the sole of a slipper. Also, a paramecium has DISTINCT FRONT AND REAR ENDS, the front rounded, the rear pointed – a good example of streamline form. And most striking of all is the rapid rate at which a paramecium swims about, as compared to the slow creeping of an amoeba.

Beneath the outer covering, and embedded in the clear outer cytoplasm, are small oval bodies called TRICHOCYSTS. These bodies reach the surface through pores and can be discharged to the exterior. During the process of discharge they become greatly elongated into fine threads.

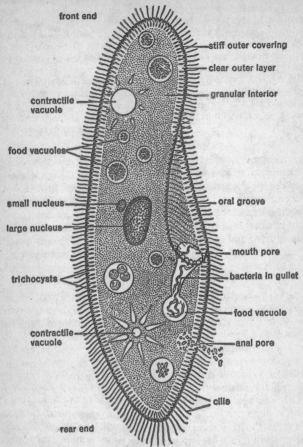

front end

stiff outer covering

clear outer layer

granular interior

contractile vacuole

food vacuoles

small nucleus

large nucleus

trichocysts

contractile vacuole

oral groove

mouth pore

bacteria in gullet

food vacuole

anal pore

cilia

rear end

A PARAMECIUM, showing principal structures. The cilia, shown only at the edge, really occur all over the surface of the body.

It is not clear what function the trichocysts serve. It is thought that they afford a means of protection, since a paramecium discharges them when touched by injurious chemicals or when attacked by an enemy. It has also been suggested that a paramecium uses the trichocysts to anchor itself while feeding on bacteria.

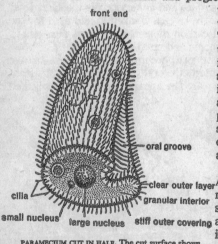

The CILIA IN A SINGLE ROW do not beat all at once but one after another, so that they appear to beat in waves. Ordinarily, they beat so fast that all we see is a flickering at the edge of the paramecium. (After Gelei.)

The paramecium has put on speed by developing accessory structures for LOCOMOTION which are not unlike the oars of a racing shell. This small animal is covered with about twenty-five hundred short 'hairs', which are really protoplasmic extensions through minute holes in the stiff surface covering. These protoplasmic extensions, called CILIA, beat in somewhat the same manner as the arms are moved in the crawl stroke in swimming; they reach forward in the relaxed part of the stroke and then give a strong backward lash. The combined effect of all the cilia, rhythmically stroking backward, is to drive the animal forward. The cilia do not beat simultaneously, but in a wave beginning at the front end of the animal and progressing backward. Further they beat obliquely rather than straight backward. The oblique stroke causes the animal to revolve on its long axis so that, as it swims through the water, it revolves continually and describes a spiral path. A paramecium can swim backward by a reversal of the ciliary stroke and can turn in any way.

The FOOD-CATCHING APPARATUS of the paramecium is much more specialized than in the amoeba. Food is taken in only at a definite place on the surface. One side of the paramecium is strongly depressed, forming a concavity, the ORAL

front end

oral groove

clear outer layer

granular interior

cilia

small nucleus   large nucleus   stiff outer covering

PARAMECIUM CUT IN HALF. The cut surface shows some of the relations that are not clear from the large diagram. The clear outer layer, containing the trichocysts, completely surrounds the granular interior. Cilia occur all over the surface of the animal.

GROOVE, as if a piece had been cut out of the animal. This concavity leads backwards to an opening, the MOUTH PORE, from which a funnel-like tube, the GULLET, extends down into the cytoplasm. When a paramecium stations itself near a bit of decaying material, the beat of the cilia in the oral groove drives bacteria and other minute organisms toward the gullet. The bacteria are whirled around by special ciliary tracts and are concentrated into a ball at the bottom of the gullet. The finished ball then passes as a FOOD VACUOLE into the cytoplasm. A paramecium that has found a suitable bit of debris and is feeding actively will soon become filled with food vacuoles. These vacuoles are moved about in the interior cytoplasm in a more or less definite course by a slow circulation of the semifluid cytoplasm, and in the meantime their contents undergo DIGESTION essentially as described for the amoeba.

The few indigestible remnants in the food vacuoles are finally ELIMINATED from the body at a definite ANAL PORE in the outer covering.

RESPIRATION and EXCRETION take place, as in the amoeba, by diffusion through the surface and are essentially the same as in all other animals – that is, oxygen is taken in and used for the burning of foods; and carbon dioxide, water, and nitrogenous wastes (said to be ammonia and urea) are given off.

Two CONTRACTILE VACUOLES occupy fixed positions near the surface on the side opposite the oral groove, one near the front end, the other near the rear. The apparatus is more complicated than in the amoeba, for each vacuole is surrounded by a circle of canals which radiate from the vacuole for some distance into the cytoplasm. At short intervals these canals fill with fluid, then discharge their contents to form the vacuole, which in turn ejects the fluid to the exterior. While the contractile vacuole of an amoeba appears to be a temporary structure

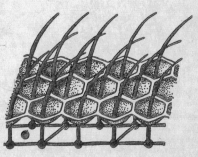

A small portion of the surface of a paramecium showing the CO-ORDINATING FIBRES connecting the bases of the cilia. (Modified after Lund.)

which re-forms before each contraction, in a paramecium the vacuole canals and the pore through which the vacuole discharges are probably permanent structures, even though the vacuole is probably not. Concerning the function of this vacuole mechanism, there is nothing to add to what was said about the amoeba. Apparently in the paramecium also the system serves primarily for regulating the water content of the animal.

In a paramecium the two contractile vacuoles can eliminate a volume of water equivalent to its body volume in about half an hour, as compared with four to thirty hours required by an amoeba. An average man eliminates a volume of urine equal to his body volume in about three weeks, but he also disposes of excess water through the lungs and sweat glands.

A cut through *Euplotes* severing the co-ordinating fibres, incoordinates the large rear cilia to which these fibres run. The cilia are shown in the figure of Euplotes on the next page. (After C. V. Taylor.)

The high degree of co-operation displayed by the cilia in the swimming movements or in food-taking suggests that some CO-ORDINATING MECHANISM resembling nervous control in higher animals is present, and such has, indeed, been discovered in the paramecium. Near the surface of the animal is a SYSTEM OF PROTOPLASMIC FIBRES (fine threads) which run longitudinally and connect the rows of small granules at the bases of the cilia. There seems little doubt that this system of fibres constitutes the mechanism which regulates the activity of the cilia; for, when the mechanism is injured, as has been done experimentally in certain ciliates, the cilia no longer beat in co-operation.

The large number of minute cilia in a paramecium makes for a co-ordinating system of fibres, or neuromotor apparatus, that is not easy to experiment upon. However, a ciliated protozoan like *Euplotes* has only a few large cilia, and it is possible to cut the fibres which run to the large rear cilia from the centre at which the co-ordinating fibres meet. After the operation these large rear cilia do not beat in co-ordination with the others, and certain of the swimming movements are interfered with. A cut of equal size, which does not injure the fibres, does not influence the co-ordination of the cilia.

The innermost granular cytoplasm, as in an amoeba, is more fluid than the surface layer, and contains food vacuoles, fat droplets, and other food bodies, as well as TWO NUCLEI, one large and one small nucleus (there are several of the small nuclei in some species of *Paramecium*). The LARGE NUCLEUS appears to be concerned with the ordinary business of the cell, while the SMALL NUCLEUS is especially active during reproductive processes.

The experimental REMOVAL OF THE SMALL NUCLEUS to determine its function is difficult in a paramecium, but can be done conveniently with *Euplotes*, which stands microsurgery well and has the added advantage

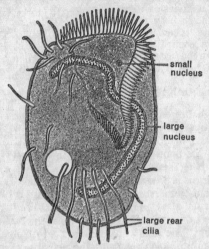

Euplotes is a good subject for experimentation. (Modified from a photomicrograph by C. V. Taylor.)

of a small nucleus that is easily visible and can be removed with a micropipette. The operated animal appears to be uninjured otherwise and may even divide once or twice; but after two or three days it dies, without ever reforming the small nucleus. That death was not due to injury other than the absence of the small nucleus is proved by the complete recovery of animals undergoing 'control' operations in which a portion of the cytoplasm, near the small nucleus, is removed, and another operation in which the small nucleus is first removed and then immediately replaced. Also, animals in which a portion of the large nucleus is removed, the small nucleus being left undisturbed, recover from the operation and produce a large number of normal descendants. It is apparent that the small nucleus, probably more so than the large nucleus, is necessary for continued life in this animal.

It is not at all clear why a paramecium (and other protozoans related to the paramecium) should have two sorts of nuclei, a condition unknown in other animals. We can only say that among these protozoans the functions of a nucleus have been subdivided between the two different bodies.

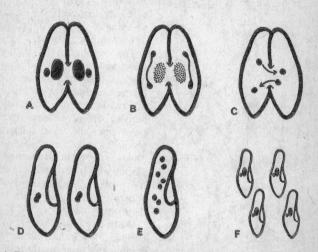

A, Two animals unite by their oral grooves. B, The large nuclei begin to degenerate; the small nuclei divide twice, three degenerate, the remaining one divides again. C, One part of each small nucleus migrates to the opposite paramecium. D, The small nuclei fuse. The animals separate. E, The product of the fusion of the small nuclei divides several times. Only one member of the pair of animals is shown. F, The animal divides twice, resulting in four small paramecia.

CONJUGATION in one species of the paramecium. Not all stages are shown. For clarity the large nuclei are omitted from C; actually they do not degenerate completely until after the conjugants have separated.

A paramecium REPRODUCES by dividing in two in a manner similar to that described for the amoeba. Both kinds of nuclei elongate and pull apart into two halves, one which remains in each daughter-cell. A constriction forms around the middle of the animal; and as the constriction deepens, the cytoplasm divides into two daughters. The front and rear halves of a paramecium are not exactly alike; but even before separation occurs, each half forms the parts necessary for a complete paramecium. Thus the gullet, which is behind the middle, falls to the rear daughter; the front daughter early in the process of division forms a new gullet.

When well fed, paramecia may divide two or three times daily, so that enormous numbers of them can be obtained in a short time. For this reason, paramecia, as well as other protozoa, have been used in many population studies. The results of such studies contribute to our understanding of the laws of growth of human populations.

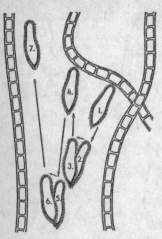

AVOIDING REACTION. 1, paramecium encounters an obstacle. 2, the animal backs. 3, shifts its position. 4, again meets resistance. 5, 6, backs and turns. 7, finds a free path. (Based on Jennings.)

The beginnings of SEXUAL PROCESSES OCCUR in the paramecium, although the animals themselves do not show visible differentiation into males and females. However, sex differences can be distinguished physiologically. Only when individuals of certain strains are placed with individuals of certain other strains do they adhere in pairs and unite by their oral grooves.

While two individuals are so united, their nuclei undergo complicated changes, the result of which is the passage of a portion of the small nucleus from each animal into the other. Each migrating nucleus fuses with the opposite remaining nucleus. The two paramecia separate and undergo a series of divisions; the animals resulting from these divisions then continue their usual activities. This sexual process is called CONJUGATION. Although more typical sexual reproduction, involving differentiated sperms and eggs, occurs in other protozoa, conjugation in paramecia has the essential features of the sexual process in all animals, that is, the *transfer of nuclear material having new hereditary possibilities from one animal to another.*

The BEHAVIOUR of a paramecium is exactly what one would expect of an animal that has no specialized sense organs to direct its movements. When not quietly feeding on bacteria, it roams about ceaselessly, bumping 'head on'. into obstacles in its path. After such a collision, the paramecium backs by reversing the beat of the cilia, turns to one side, and goes off in a new direction. If this second path results in another collision with the same obstacle, the set of movements is repeated. Finally the animal encounters a free path and continues

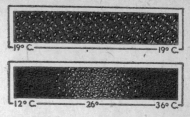

BEHAVIOUR IN RELATION TO TEMPERATURE. When the temperature is uniform throughout the tank, the paramecia are uniformly distributed. If one end of the tank is cooled to 12°C. and the other end is heated to 36°C., the paramecia avoid the extremes of temperature and accumulate in the region of optimum (most favourable) temperature. (After Mendelssohn.)

on its course. The set of movements with which a paramecium backs, turns, and swims off in a new direction is called the AVOIDING REACTION. Mechanical obstacles, excessive heat, excessive cold, irritating chemicals, unsuitable food, a predaceous enemy – all elicit the avoiding reaction, which may be said to constitute most of the behaviour of a paramecium.

In its constant explorations the paramecium may swim by chance into a region rich in bacteria. Each time that it crosses the boundary of this region into a less favourable area, it gives the avoiding reaction; thus it remains in the more favourable region. This general method of finding the best conditions of existence is called TRIAL-AND-ERROR BEHAVIOUR and is employed to some extent by all animals, including man.

A paramecium need not actually enter an unfavourable region before it can react negatively. The beating of the cilia in the oral

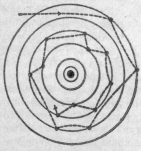

TRIAL-AND-ERROR BEHAVIOUR IN RELATION TO CHEMICALS. A small amount of a chemical is placed in the centre of a drop of water, and it slowly diffuses outwards. The concentric circles indicate zones of diminishing concentration of the chemical from the centre. The broken line shows the path of a paramecium placed in the drop. As the animal swims about, it gives an avoiding reaction whenever it enters a zone less favourable than the one it is in; it thus appears to be 'trapped' in the most favourable region. (After Kuhn.)

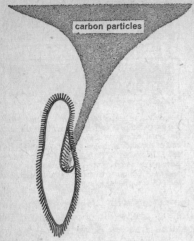

carbon particles

PARAMECIUM 'SAMPLES' THE WATER AHEAD, as can be seen by placing in the path of the animal a drop of Indian ink containing visible particles. (After Jennings.)

groove draws a constant stream of water, in the form of a cone, towards the oral groove. If there is an irritating chemical in the water ahead or if the water is hotter or colder, a portion of water from the new region will be drawn backward into the oral groove. Thus the paramecium constantly receives 'advance information' of the environment ahead and responds with the ˉavoiding reaction without actually entering an unfavourable region.

Paramecia have only a poorly developed ability to discriminate between foods, since they very readily take in and form food balls of almost any minute particles, such as carbon grains, dye suspensions, and the like. However, after a time they will reject these inert particles while still accepting bacteria. Paramecia avoid strong acids; but they give the avoiding reaction when passing from dilute acids to ordinary water, and therefore tend to aggregate in regions of low acidity. This behaviour aids the animal in feeding, because bacteria are most likely to be present near decaying organic matter, which renders the surrounding water slightly acid. On the whole, it may be said that the behaviour of a paramecium is remarkably adaptive for an animal that has to find its way about simply by keeping out of trouble.

CHAPTER 4

Classified Knowledge

To the casual observer porpoises and sharks are kinds of fish. They are streamlined, good swimmers, and live in the sea. To the zoologist who examines these animals more closely, the shark has gills, cold blood, and scales; the porpoise has lungs, warm blood, and hair. The porpoise is fundamentally more like man than like the shark and belongs, with man, to the mammals – a group that nurses its young with milk. Having decided that the porpoise is a mammal, the zoologist can, without further examination, predict that the animal will have a four-chambered heart, bones of a particular type, and a certain general pattern of nerves and blood vessels. Without using a microscope he can say with reasonable confidence that the red blood cells in the blood of the porpoise will lack nuclei. This ability to generalize about animal structure depends upon a system for organizing the vast amount of knowledge about animals.

This knowledge was not always so well classified. Only a few hundred years ago the study of animals was in a crude descriptive stage. The accumulations of facts about animal structure and animal behaviour were almost entirely useless because little attempt had been made to relate one fact to another and because the facts were arranged in categories based on superficial distinctions. Biological science made little progress until it was realized that animals must be grouped according to their fundamental similarities in structure and then arranged into a workable SYSTEM OF CLASSIFICATION.

The heterogeneous assortment of organisms that compose the animal kingdom are first divided into large groups, called PHYLA, which are based on *radically different plans of organization*. The members of a phylum may live in every kind of habitat, may vary in size and body form, and in their methods of locomotion and feeding – but they have a common basic structure. In the Protozoa this underlying structure is the differentiation of protoplasm within a single cell – other phyla have other body plans. The chief phyla are well recognized and agreed upon by zoologists, though there is debate over the classification of some of the rarer and more highly specialized or degenerate animals. Unfortunately, there is always a certain amount of arbitrariness in any system of classification, and the exact number of phyla that we name depends upon our criteria of a radically different plan of organization.

Within each phylum the members are further divided into groups, called CLASSES, on the basis of a *significant variation* in the fundamental plan, usually in adaptation to a *special way of life*. To take a crude analogy from everyday experience: if we regard all vehicles driven by gasoline engines as belonging to the same category or 'phylum', significant variations are automobiles, aeroplanes, and motor boats. Each vehicle has the same fundamental plan, a gasoline engine – yet each is constructed for a significantly different kind of travel. Similarly, protozoa that move by pseudopods, like the amoeba, are grouped in a separate class from protozoa that move by means of cilia, like the paramecium.

Each class is further divided into smaller categories called ORDERS. In terms of the analogy given above, we can subdivide the class automobiles into 'orders': passenger cars, trucks, racing-cars. Order differences are still of such magnitude that they can be recognized easily. For instance, the class Insecta has orders such as the beetles, the flies, the butterflies and moths, the fleas, and others.

Each order consist of a number of FAMILIES. The anatomical distinctions between familes are still important enough to be of survival value to the groups concerned. That is, the structures which serve as a basis of classification are likely to be the ones that enable a member of one family to live in a place that is uninhabitable for a member of another family. A diving beetle that has its legs modified for swimming and its jaws for seizing prey would have difficulty getting along in a forest, the home of a leaf-eating beetle.

The anatomical criteria used to divide a family into groups called GENERA (singular, genus) are usually so small that they would not be noticed by most people and in general have less adaptive value for the animal concerned than have family distinctions. Crayfish of the genus *Cambarus* have seventeen pairs of gills, while those of the genus *Astacus* all have eighteen pairs of gills. The possession of the extra pair of gills probably makes no difference in the success with which *Astacus* meets its environment as compared to *Cambarus*. *Astacus*, which is found only west of the American continental divide, would probably do well east of the Rockies in the territory of *Cambarus*. But the two groups of crayfish have long been separated by an impassable barrier, the Rocky Mountains.

When we divide a genus into SPECIES, we are at last dealing with a category which is somewhat less arbitrary and which represents what the scientist means by a *kind of animal*. There are borderline cases that do not clearly fit this definition; but for the vast majority of animals a species may be defined as a natural population of organisms which has a heredity distinct from that of any other group, and the members of which breed only with one another to produce fertile offspring. To return to the example of the crayfishes, the genus *Cambarus* can be divided into distinct species on the basis of minor details, particularly the shape of the first pair of appendages on the abdomen of the male. One of these species lives in rivers and is called *Cambarus limosus*. Another, called *Cambarus diogenes*, lives in swamps, where it burrows in the mud. The minor anatomical difference in the appendages of the male, by which we can distinguish the two, has no practical bearing on the more deep-seated physiological differences which really make these crayfishes two species or populations that do not breed together. Physiological differences are not easy to measure or define, and they are difficult or impossible to determine from dead specimens. Since animals must often be classified long after they are collected, it is desirable, whenever possible, to base the criteria for identification of the species on some easily observable anatomical character which is not changed by the

SPECIES DIFFERENCES in two crayfish – the first abdominal appendage of the male. *Cambarus limosus*, left; *C. diogenes*, right. (After Ortmann.)

death of the animal. Thus, a few distinctive characters are selected for identification of species. However, the individuals of two species differ in not one but a great many minor characteristics of form and behaviour, and these differences are due to the accumulations of minor changes in the course of evolution.

Species cannot be described in all cases by clearly visible anatomical differences; sometimes physiological characteristics must be studied. For example, two species of a single-celled green alga (*Chlorella*) can be distinguished only by measuring their average rates of respiration. There are two species of termites that can be distinguished most easily by identifying the other animals that live associated with them. Except for this difference in the 'guests' which they harbour, the termites can be distinguished only on the basis of average differences in certain body measurements made on a large number of individuals. A single termite, found away from the nest, cannot be assigned definitely to either species. Such cases emphasize the need for dealing with populations, rather than with single individuals, in describing a species.

In a widespread species there may be a gradual change in the characteristics of the population from one end of its realm to the other, so that in some instances widely-separated individuals from the same population would be regarded as different species if there were no intergrades. When there are intergrades, such widely separated forms are regarded as belonging to different SUBSPECIES.

If we refine our observations and criteria, we can often distinguish, among the members of a species, individuals that can, but usually do not, breed together or live together in the same region. They exhibit minor differences, such as the colour variations in the skin of man. The differences are usually more superficial than those which distinguish a species, and the groups classified on this basis are termed VARIETIES, RACES, or STRAINS. Still less significant are differences between the individuals of a variety or race.

The double name by which we referred above to the river and swamp crayfishes is called the SCIENTIFIC NAME and consists of, first, the genus name (written with a capitalized initial letter) and, second, the species name (not capitalized). Man's scientific name is *Homo sapiens*. A common paramecium is *Paramecium caudatum*. The name *caudatum* means tailed and refers to the long cilia at the rear. *Paramecium caudatum* is distinguished from *Paramecium aurelia* by the possession of one small nucleus instead of two, the more pointed rear end, and larger body size.

The scientific name is frequently written with the name of the scientist who first adequately described and named the species. For example. '*Balanus balanoides* Linnaeus' is the common barnacle, which was first described by Linnaeus (Carl von Linné), a Swedish naturalist who established the system of giving two names to animals. In his time (1707-78) Latin and Greek were the accepted languages for scientific writings, and Linnaeus gave all the animals Latin names. This

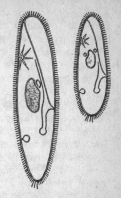

TWO SPECIES OF PARAMECIA. *Paramecium caudatum*, left; *Paramecium aurelia*, right. (After Wenrich.)

system has become universally adopted, and it is now obligatory for scientists to produce a Latinized name for a newly-described animal. Almost a million animal species have already been described and named by this system, and it is probable that several times this number are still to be discovered. Every year several thousand new species are added to the list of catalogued animals.

When a zoologist mentions an animal in a scientific paper, he does not use its common name, because this varies from place to place and from time to time. What is loosely called a 'crayfish' in one locality may be an entirely different species or a different genus from what is known by the same common name in another locality. The scientific name is international, recognized the world over as referring only to one clearly-defined species of animal. Sometimes a species is so distinctive that it can be recognized at sight or by reference to a catalogue. But, as a rule, the number of similar species is so great and the differences between them so small that they can be accurately identified only by a specialist in classification, a taxonomist, who knows the detailed characteristics of the group. There is no virtue in giving the full name of an animal if the name is not correct. Rather than commit this 'scientific crime', one refers to a common animal by using its genus (or generic) name. Thus we have said, in chapter 3, that one could operate on a *Euplotes* as if it were a particular animal. Yet the name actually refers to a genus consisting of several species of animals. Sometimes we do not even capitalize the generic name, if the animal to which it refers is so well known that its generic name has become a common name. We may refer to an individual of any species of the genus *Paramecium* simply as a 'paramecium'.

THE system of classification represents a SCHEME OF ANIMAL RELATIONSHIPS. When we see two men who are strikingly similar, we are likely to say: 'They must be brothers.' If they resemble each other somewhat less, we may say: 'No doubt they are cousins.' If they bear the same name but resemble each other hardly at all, we guess that they are only distant relatives. In other words, we judge the closeness of their relationship by their degree of similarity. Two brothers have a pair of very recent common ancestors, their parents. Two cousins have only a pair of grandparents in common, a less recent common origin. In the same way, two species of the same genus have had a fairly recent common ancestor. Species of two different genera have had a more remote common ancestor. And species of two different families are still more distantly related. Thus, the position of an animal in the scheme of classification indicates our idea of its relationship to other animals.

CHAPTER 5

A Variety of Protozoa

PROTOZOA are cosmopolitan – the same species may be found on every continent. In sharp contrast with the provincial habits of most larger animals, which are limited in their spread by ocean or land barriers, protozoa are readily swept along in ocean or river currents, or in the encysted state may be blown by the wind or transferred from one pond to another in the mud that clings to the feet of birds. Encystment not only furthers the distribution of protozoa but also enables them to live in habitats they otherwise could not invade. Within the heavy walls of their cysts, some protozoa can resist the heat and drought of summer in the desert. After the first rain the cysts break open and the protozoa move about, frequently encysting again after only an hour of activity. Animals of such versatility are well adapted to exploit the rich possibilities offered by the moist and nutritious interiors of other animals. Man is a favoured host, but protozoan parasites are rather impartially distributed among all animals.

About fifteen thousand species of protozoa have already been described, though it is probable that many more remain to be discovered. Several other phyla have many more species; but when we consider that all larger animals, and even some protozoa, harbour one or more species of protozoa, we are led to the interesting conclusion that there are many more individual protozoa than individuals of all other animals combined.

The flagellates and other protozoa are sometimes placed with the bacteria in one group known as the PROTISTA. This device eliminates

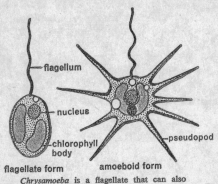

flagellum

nucleus

chlorophyll
body

pseudopod

flagellate form   **amoeboid form**

Chrysamoeba is a flagellate that can also
feed like an amoeba. (Modified from Lang.)

the difficulty of distinguishing primitive plants from primitive animals and takes care of the many forms intermediate between protozoa and bacteria, such as the spirochetes (causative agents of syphilis). On the other hand, some zoologists separate these groups, leaving not only bacteria but also all green flagellates to the botanists. We shall take an intermediate view, excluding bacteria and spirochetes, but including green flagellates among the members of the first phylum of animals.

The phylum PROTOZOA may be conveniently divided, according to their chief method of locomotion, into five classes: the flagellates, the amoeboid protozoa, the spore-formers, the ciliates, and the suctorians.

THE FLAGELLATES

THE flagellates (class FLAGELLATA) are protozoa that have one or more long filamentous protoplasmic extensions, FLAGELLA, by which they swim. They have a more or less definite shape, usually oval, and a definite front end from which the flagella arise. The flagella beat in whiplike fashion, and the locomotion of flagellates is usually slower and more irregular than that of a paramecium and its ciliated relatives.

The flagellates are divided into two groups, the more primitive PLANT-LIKE types which make their own food by means of the chlorophyll they possess, and the ANIMAL-LIKE types which capture and ingest other organisms.

One of the simplest of the plantlike flagellates is *Chrysamoeba*, which has a yellow photosynthetic pigment. This form sometimes temporarily loses its single flagellum. It then moves about and ingests solid food by means of pseudopods, like an amoeba. Since such flagellates are both independent and capable of feeding like animals, they are regarded as being more primitive than the amoebas, which never show flagellate stages and feed only like animals.

Euglena is one of the most common of the green flagellates of fresh waters and is often so numerous that it produces a green scum on the surface of ponds. The name applies to a number of species, all of which have a very elastic outer covering and specialized protoplasmic contractile fibres which permit contraction and elongation of the body in a characteristic squirming called EUGLENOID MOVEMENT. At the front end there is a flask-shaped depression, the gullet, from which springs one very long delicate flagellum. By lashing this flagellum, the animal swims slowly forward, at the same time revolving on its long axis, as does a paramecium. Since Euglena has never been seen to feed, it is probable that the gullet is not used to take in food but serves mainly for the attachment of the flagellum. A large contractile vacuole discharges its contents into the gullet at frequent intervals.

EUGLENOID MOVEMENT. 1–5, a succession of characteristic shapes.

Beside the gullet is a bright-red granule,

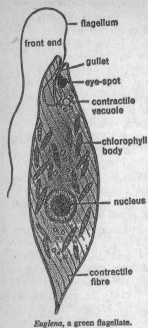

Euglena, a green flagellate.
(Modified after Doflein.)

called the EYESPOT, which is thought to be associated with a light-sensitive apparatus. The specialized eyespot of Euglena is more sensitive and therefore more efficient as a light-detecting apparatus than the undifferentiated protoplasm of the amoeba, which also reacts to light. This specialization is important, since euglenas depend upon photosynthesis for their food, and it is advantageous that they should expose themselves to the light as much as possible. When placed in a dish, euglenas quickly aggregate on the side of the dish nearest the light. As long as they are exposed to light, they maintain themselves by photosynthesis and store up carbohydrates in storage bodies which are conspicuous in the cytoplasm. However, euglenas can live in the dark if they are placed in a nutrient solution. Under these conditions the chlorophyll degenerates, the animals become colourless, and the nutrient materials are absorbed through the surface membrane. It is doubtful if this mode of nutrition is ever used in nature.

Volvox is seen in fresh-water ponds as a small, green sphere which may be one-tenth of an inch in diameter. The sphere is composed of thousands of flagellates embedded in the surface of a jelly ball. Each flagellate resembles Euglena and has two flagella, a red eyespot, two contractile vacuoles, chlorophyll bodies, but no gullet. Volvox is a COLONY of unicellular animals rather than a many-celled animal, because even the simplest many-celled animals have considerably more differentiation between cells than appears among the cells of Volvox. The colony swims about, rolling over and over from the action of the flagella; but, remarkably enough, the same end of the sphere is always directed forward, and thus we can distinguish front and rear ends. Its behaviour can be explained only by supposing that the activities of the numerous flagellates are subordinated to the activity of the colony as a whole. If the flagella of each member of the colony were to beat without reference to the other members, the sphere would never get anywhere. In such subordination of the individual cells of a colony to the good of the colony as a whole we see the BEGINNINGS OF INDIVIDUALITY as it exists in the higher animals, where each animal behaves as a single individual, although composed of millions of cells.

The CO-ORDINATION of the Volvox members in swimming implies some means of transmission between them. They are, in fact, connected by protoplasmic strands extending through the jelly. The co-ordination of numerous components into an

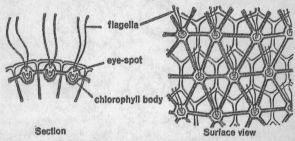

Section   Surface view

Volvox. The cells of the colony are joined by protoplasmic connections. (Modified after Janet.)

individual is usually followed by the specialization of different individuals for different duties. Only the slightest degree of specialization is seen in the Volvox colony; the flagellates of the back part of the colony are capable of reproduction, while the front members never reproduce but have larger eyespots and serve primarily in directing the course of the colony.

In ASEXUAL REPRODUCTION (that is, not involving any sexual processes) a cell enlarges, loses its flagella, and divides a number of times until a small daughter-ball is produced. A Volvox sphere usually contains in its hollow interior several such daughter-colonies in process of formation. They are liberated when the mother-colony breaks open.

In the SEXUAL REPRODUCTION of Volvox one of the stages in the evolution of sex is illustrated. In the early stages of sexual differentiation the two fusing cells are alike, and such a condition is seen in some of the relatives of Volvox. In Volvox this evolution has reached completion; and fully differentiated male gametes, or SPERMS, and female gametes, or EGGS are formed. In forming an egg, a cell of the colony increases greatly in size, takes on a rounded form, and becomes loaded with food, especially fatty substances. This food is contributed in part from adjacent cells and serves to give the young Volvox a good start in life. Another cell of the same or another colony, by repeated divisions, gives rise to numerous small flagellated sperms. These sperms swim about until they find an egg or die. Only one sperm penetrates the egg, and this union of eggs and sperm is called FERTILIZATION. When an egg is fertilized, it

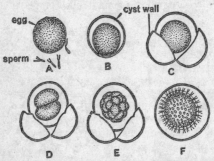

SEXUAL REPRODUCTION AND DEVELOPMENT OF *Volvox*. A, fertilization. B, the zygote forms a cyst wall. C, an inner cyst wall forms, and the outer wall is discarded. D, the zygote divides into two cells. E, a small ball of cells. F, a young colony within the cyst wall. (Modified after Kircher.)

undergoes a change, which prevents the entrance into the egg of additional sperms. The fertilized egg, or ZYGOTE, of Volvox secretes upon its outer surface a hard spiny shell which protects the cell during unfavourable conditions, such as drying or freezing. Inside the shell, the zygote divides into two cells, these into four, and so on until a small ball has been produced. With the return of favourable conditions of heat and moisture, the shell breaks; and the young colony, indistinguishable from a daughter-colony produced asexually, emerges.

The DINOFLAGELLATES occur in enormous numbers in the surface waters of the ocean; there are also a number of fresh-water species. A typical dinoflagellate, *Gonyaulax*, is enclosed in a tight-fitting armour of cellulose plates and has two flagella, one lying in a grove encircling the animal and the other trailing downward. Many dinoflagellates have oil drops which help them to float. Some of them possess a brown pigment in addition to chlorophyll; others are colourless and feed on minute organisms. Some dinoflagellates produce light, as do many other marine and some land animals. The mechanism of light production in certain of the higher forms has been partially worked out (and will be given in chap. 18), but has not been proved to apply to these protozoa. One of the most abundant of the luminescent dinoflagellates is *Noctiluca*, which does not resemble the typical dinoflagellates, but is colourless and ingests solid food. It is spherical, about a millimetre in diameter, with one short thick and one very delicate flagellum. It floats on the surface near shores, often in inconceivable numbers, and, being faintly pink, may in the daytime cause whole

A DINOFLAGELLATE, *Gonyaulax*, sometimes becomes so abundant along the coast of southern California that it colours the ocean red for miles. These outbreaks of 'red water' cause the death of fish and other animals, which are cast up on the beach and decay. The stench has been compared to that of the Nile when, according to ancient writings, that river 'turned to blood'. (After Kofoid.)

areas of the ocean to look like weak tomato soup. The minute flashes emitted by these dinoflagellates are seen at night when the animals are agitated as the waves strike rocks or other objects.

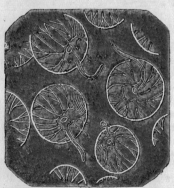

A LUMINESCENT DINOFLAGELLATE, *Noctiluca*. (Based on several authors.)

Night swimming or boating in an area filled with Noctiluca causes these dinoflagellates to emit light, furnishing spectacular displays of what appears like fireworks being shot off under water.

Among the animal-like flagellates are the COLLAR-FLAGELLATES, solitary cells that live attached by their stalks to the substratum. They become colonial when the cells fail to separate completely after dividing.

The cell has a delicate, transparent protoplasmic collar from the centre of which emerges the single flagellum. The beating of the flagellum draws a current of water toward the cell. The food organisms in the current do not enter the collar but impinge upon the sides of the cell and are taken up into food vacuoles. The collar-flagellates are of special interest because a similar type of cell occurs nowhere else in the animal kingdom except in the sponges.

Many parasitic forms are included among the animal-like flagellates. Of these, the most notorious are the TRYPANOSOMES

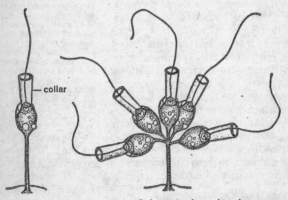

A COLLAR-FLAGELLATE, *Codosiga*. A colony arises when the cells fail to separate after division. (After Lapage.)

that cause African sleeping-sickness in man (not the same as the sleeping-sickness known as 'encephalitis' and caused by a virus). The trypanosomes and their relatives live as parasites in insects, certain plants, and most vertebrates in Africa without causing much inconvenience to their hosts. But apparently they have not yet become adapted to living in man or in his domestic animals without producing an incapacitating and usually fatal disease.

The trypanosomes that cause African sleeping-sickness are transmitted by the blood-sucking TSETSE FLIES. Practically all wild game in Africa harbour trypanosomes in their blood; and when a tsetse fly sucks the blood of an antelope, for example, or of an infected man, the blood drawn into the intestine of the fly will contain these trypanosomes. In the fly's intestine the trypanosomes undergo changes. From the intestine they invade the salivary glands of the insect, where they continue to change and to multiply. If such an infected fly bites a man, the trypanosomes are injected into the blood of the victim with the saliva of the fly. In the blood they multiply rapidly and wriggle about among the blood corpuscles, propelled by the undulations of a delicate ruffled membrane which extends along one side of the body.

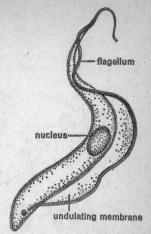

A TRYPANOSOME, the cause of African sleeping-sickness. The edge of the undulating membrane bears the flagellum, which extends free at one end of the animal.

The bite of an infected fly sometimes does not cause fever for weeks or even months, while the flagellates increase in number. When the attacks of fever begin, the victim becomes weak and anaemic, probably because of poisonous by-products of metabolism given off by the flagellates. Finally, the parasites invade the fluid surrounding the brain and spinal chord, the person loses consciousness, and the 'sleeping-sickness' goes to a fatal end.

Medical treatment of African sleeping-sickness consists, in the main, of injection of various drugs, and is very effective in the early stages of the disease before the flagellates have invaded the nervous system. However, if the dose injected is not adequate to kill the parasites, they may become drug-resistant and then are unaffected by doses large enough to kill the patient.

Control measures for this disease are complicated by the fact that even if it were possible to cure or isolate all human cases, the wild game would still act as a reservoir from which flies could be reinfected. An adequate solution for this problem seems to lie in wholesale destruction of the tsetse flies, which is an exceedingly difficult task, although not impossible.

Large regions in Africa are uninhabitable for men and their stock because of the tsetse fly, and it is safe to say that the past and present accomplishments of man in the larger part of Africa have been controlled by the fly. The future of that continent depends not upon military leaders or diplomats of state but upon medical and ecological investigators who are working to check the ravages of that murderous pair, the fly and the flagellate.

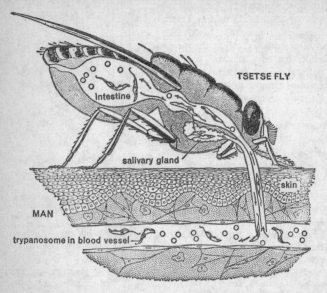

The LIFE-CYCLE OF THE TRYPANOSOMES that cause African sleeping-sickness involves two hosts, the tsetse fly and a vertebrate, such as an antelope or a man. Here an infected fly is biting an infected man. Principal stages in the life-cycle are shown: trypanosomes in the salivary glands of the fly being injected into the blood of the man, forms living in the blood and entering the sucking tube of the fly, and those living in the intestine of the fly. (Based on several sources.)

The TRICHOMONADS are flagellates common in the digestive tracts of vertebrates. They are pear-shaped protozoans with several flagella springing from the anterior end. One of these flagella extends backwards along the edge of an undulating membrane. The body is supported by an internal stiff rod which projects from the rear end and is frequently used to anchor the animal while it feeds. *Trichomonas buccalis* inhabits the mouth and may be

Shaded portion shows the extent of sleeping-sickness in Africa.

involved in pyorrheal conditions; *Trichomonas hominis* and some other kinds of trichomonads live in the intestine and seem to be related to a type of diarrhoea.

The flagellates are considered to be the simplest class of protozoa because certain of their members are very primitive. It must be realized, however, that no class of animals is made up only of simple forms. Practically all the large groups have a few members which are more complex than the less developed members of the next higher group. The older an animal group, the longer has been its evolution and the time available for development of complexity. It is not surprising, then, to find certain flagellates that show an amazing degree of specialization. One of these, *Trichonympha*, is among the most complex of all protozoa. It is also interesting because it inhabits the intestine of wood-eating termites. It was for a long time difficult to understand how the termites were able to subsist on a wood diet, since wood contains only a very small amount of digestible protein and sugar, and most animals lack the necessary enzymes

Trichomonas vaginalis is found in the vagina (passage leading from the uterus to the exterior) and is under suspicion as a contributing factor in death from childbirth. (After Powell.)

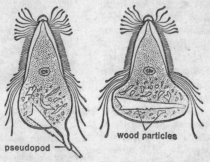

pseudopod

wood particles

Trichonympha engulfs minute bits of wood by pseudopods extended from the lower part of the animal. This method of ingestion is surprising, in view of the fact that the surface cytoplasm of the upper part of the animal is covered with hundreds of long flagella and is structurally more complex than in almost any other protozoan. (Modified after Swezy.)

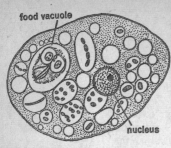

A HARMLESS AMOEBA, *Entamoeba coli*, lives in the intestine of man and feeds on particles in the intestinal contents. This particular specimen has just rendered a useful service to its host by engulfing a harmful parasitic flagellate called *Giardia*. The smaller food vacuoles contain bacteria. (After Doflein.)

for digesting its chief constituent, cellulose. In recent years evidence has been presented to show that flagellates, such as Trichonympha, which lives sheltered in the termite's intestine, ingest the minute bits of wood occurring there and transform them to soluble carbohydrates, part of which can be utilized by the insect host. This relationship, in which the members of two species live together in an association of mutual benefit, is an outstanding example of MUTUALISM (or symbiosis).

THE AMOEBOID PROTOZOA

THE amoeboid protozoa (class SARCODINA) include all those protozoa which move about and capture food by means of pseudopods. Certain of these resemble closely the typical amoeba, described in chapter 3, and live free in fresh and salt waters and in damp soil.

In addition to the free-living amoebas, there are a number of species which live in the interior of animals, particularly the digestive tract. Most of them are harmless to the animal they inhabit, living in the intestine and feeding on bacteria and food fragments, at no expense to their host – a relationship known as COMMENSALISM. A few species are definitely parasitic, that is they attack the host itself and frequently cause disease. These parasitic amoebas resemble in appearance and activities the free-living

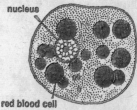

THE DYSENTERY AMOEBA, *Entamoeba hystolytica*, causes occasional epidemics in temperate regions, as when faulty plumbing allows sewage to get into the water supply. In tropical countries amoebic dysentery is a constant menace and a serious drain on the energy of the natives. Travellers in the Orient, where human faeces are used as fertilizer, should never eat uncooked vegetables or drink unboiled water. (After Doflein.)

species, but lack contractile vacuoles.

About half a dozen species of amoebas may live in man, of which only one, the DYSENTERY AMOEBA, *Entamoeba histolytica*, is definitely harmful. The dysentery amoeba occurs in 5–10 per cent of the population of civilized countries and in as much as 60 per cent of the people in backward communities. It is passed on from one person to another by eating food or drinking water contaminated with the excreta of individuals already infected with the amoeba. The dysentery amoeba inhabits the large intestine, where it feeds upon the living cells and tissues, leading to the formation of abscesses and bleeding ulcers. The active amoebas cannot live outside the body and there do not serve to

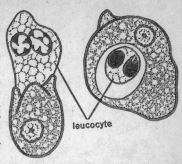

leucocyte

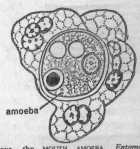

amoeba

Above, the MOUTH AMOEBA, *Entamoeba gingivalis*, engulfs a leucocyte. *Below*, the tables are turned; three leucocytes have joined forces and are engulfing an amoeba. (Modified after H. Child.)

transmit the disease, but they pass readily into the encysted stage, in which the nucleus divides twice to form four nuclei. These four-nucleated cysts pass out in the stools and, when swallowed by other persons, infect them with the disease. Various drugs, containing iodine, such as diiodohydroxyquin, are used in the treatment. It seems to be specific that it is deadly for *Entamoeba histolytica* but not for the other intestinal amoebas. Parasitic amoebas, like most disease-producing organisms, are better avoided than killed. Sanitary disposal of faecal matter and cleanliness in preparing food are essential in the prevention of the disease.

The MOUTH AMOEBA (*Entamoeba gingivalis*) is found in a large part of the human population (perhaps 75 per cent or more) in the mouth, where it feeds on bacteria and loose cells. These

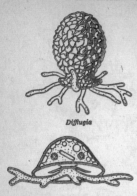

Difflugia

Arcella is enclosed by its hemispherical shell. The pseudopods extend from the hole on the under side.

amoebas occur chiefly in the pockets formed between the teeth and the gums in the disease pyorrhea. It is not certain that the amoebas play any role in the initiation of pyorrhea, but it is probable that they aggravate the disease, once it is started. This amoeba is not known to form cysts and is spread from mouth to mouth in eating and in kissing. The best one can do to reduce the probability of infection is to take ordinary care of the mouth.

Difflugia is a free-living amoeba which gathers sand grains, cements them together with a sticky secretion, and thus builds a kind of house which it carries about with itself and into which it withdraws when disturbed. The various species of Difflugia can be recognized by the specific shapes of the protective coverings which they construct. Thus, while we appear to be classifying these amoebas on the basis of differences in structure, we are also classifying them by their differences in behaviour. The common *Arcella* of fresh-water ponds secretes a hard shell about itself. When the animal divides, one daughter retains the shell; the other has to construct a new one.

The FORAMINIFERS ('hole-bearers') are amoeboid protozoans that secrete many-chambered calcareous (chalky) shells. They occur abundantly in marine waters, some species floating near the surface, but most of them living on the

In a LIVING FORAMINIFER the pseudopods extend through the pores in the walls of the shell as well as out of the main opening.

mud of the ocean bottom. A young foraminifer resembles an amoeba and secretes a shell about itself. As growth continues, the protoplasm flows out of the shell opening, spreads over the surface of the shell, and secretes another shell – thus making a second chamber. This process is repeated as the foraminifer grows, until sometimes there are more than a hundred communicating chambers. In many common species the chambers are added on in a spiral pattern, resulting in a shell that resembles a miniature snail shell. The animal occupies all the chambers and sends out long,

The skeleton of a RADIOLARIAN may be relatively simple, with only a few large spines like this one, or may be an almost unbelievably intricate latticework of silica. (After Haeckel.)

delicate pseudopods through the pores in the walls of the shell, as well as out of the main opening. The pseudopods exhibit constant streaming movements and unite into pseudopodal networks in which food is captured and digested.

Apart from their architectural accomplishments the foraminifers are remarkable among protozoa for the large size that some species attain. Living species may be larger than a pinhead, and certain extinct fossil foraminifers were as large as a sixpence. However, most forms are just visible to the naked eye.

It is astounding to contemplate the probable number of individual shells composing the chalk cliffs of Dover or the large chalk beds (1,000 feet thick in places) of Mississippi and Georgia. The presence of chalk beds indicates that these regions were once covered by the sea, and such information is useful to geologists. The deposition of the shells, if we are to judge by the present rate of accumulation at the bottom of the Pacific, was about 2 feet in a hundred years. Most of the shells now being deposited are those of the foraminifer *Globigerina*, which floats in the surface waters. These animals are constantly dying and their skeletons sinking in a slow but steady rain to the ocean floor, where they form a grey mud called 'globigerina ooze'. About 30 per cent of the floor of the ocean (40,000,000 square miles) is covered with globigerina ooze. Some deposits form chalk, with as much as 90 per cent calcium carbonate. Nearer the shore the deposits contain sediments which have been washed from the land, and the resulting rock is a fossiliferous limestone such as the famous Indiana building-stone.

Extinct foraminifers preserved as fossils in rocks are of great value in developing oil fields. Borings at different depths are examined for their fossil foraminifers, and from the species present a great deal can be learned about the underlying rock structure.

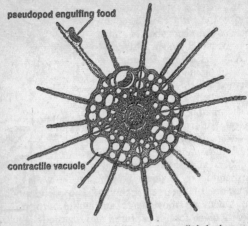

A common HELIOZOAN, *Actinophrys sol*, is sometimes called the 'sun animalcule' because its stiff radiating pseudopods, whose firm protoplasmic axes run through the cytoplasm and converge around the centrally located nucleus, suggest the rays of light from the sun.

The RADIOLARIANS are amoeboid protozoans that secrete elaborate skeletons composed mostly of *silica*. They extract the silica from the sea water just as foraminifers extract calcium carbonate. Food organisms are caught on the stiff pseudopods that radiate out through holes in the skeleton. Like many other marine protozoa, they have no contractile vacuoles. The cytoplasm is filled with (noncontractile) vacuoles that give it a frothy appearance and enable the animals to float.

Radiolarian skeletons can be found in marine deposits in shallow water; but it is only in very deep regions that they occur in a concentration of at least 20 per cent and the deposit can be classed as 'radiolarian ooze'. This ooze occurs in the Pacific and Indian oceans, where it covers an area of almost 3,000,000 square miles. Hardened radiolarian deposits are found in other rocks as siliceous inclusions, flint or chert.

The HELIOZOANS are the only fresh-water amoeboid protozoa which are comparable to the exclusively marine radiolarians.

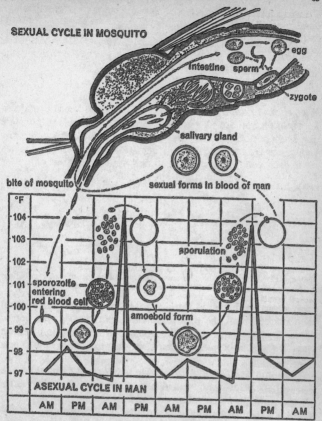

Diagram of the LIFE-CYCLE OF THE MALARIAL PARASITE. The phases of the cycle in the mosquito are shown in the diagram of the mosquito. The phases in the blood stream of man are shown superimposed on the temperature chart of a malaria patient. The bursting of millions of red blood cells, liberating spores and black granules, coincides with the regular periods of high temperature. (After various sources.)

Some forms have perforated siliceous skeletons which resemble those of radiolarians, but others have only a gelatinous covering or a skeleton of loosely matted needles of silica. Fresh-water species have a contractile vacuole. The radiating pseudopods are even stiffer than those of the radiolarians and show little movement besides a streaming of the granules in the protoplasm. When a large organism is being engulfed, several pseudopods work together. The firm protoplasmic axes of the pseudopods are absorbed, and the pseudopods wrap themselves around the prey and inclose it in a food vacuole.

THE SPORE-FORMERS

THE spore-formers (class SPOROZOA) have no special mode of locomotion and are all parasites. Some of them are responsible for human disease. They are called sporozoans because they reproduce by 'spores', in the formation of which the nucleus of a protozoan divides many times until a number of nuclei

Anopheles   Culex

Malaria is transmitted to man only by mosquitoes of the genus *Anopheles*, whose characteristic attitude while biting is shown contrasted with that of *Culex*, one of our most common mosquitoes. *Culex* transmits the sporozoan that causes bird malaria.

have been produced. A little cytoplasm then gathers around each nucleus, and the protozoan falls apart into a number of offspring, corresponding to the number of nuclei. The products of such sporulation may be naked or they may be enclosed in a resistant wall.

An example of the Sporozoa is the MALARIAL PARASITE, the cause of malarial fever in man and other warm-blooded vertebrates. This parasite enters the red blood cell as a minute amoeboid form which feeds and grows until it occupies almost the entire volume of the cell. After attaining full size, the parasite undergoes sporulation, dividing into a number of offspring. The cell breaks up, and the young spores and black granules (waste products from the decomposition of the red substance, haemoglobin, in the cells) are set free into the circulating blood of the infected person, causing an attack of chills and fever. The offspring next penetrate into new cells and repeat the foregoing history, sporulating in turn into new batches of offspring. In this manner there are soon produced millions of parasites, which destroy a large percentage of the red blood cells. After asexual multiplication has continued for some time, certain of the parasites develop into sexual forms. These develop no further unless blood containing them is sucked up by a mosquito of the genus *Anopheles* which happens to bite an infected person. Within the stomach of the mosquito, the egg-producing cell becomes an egg while the sperm-producing cell divides into sperms, which then escape. The sperms are very slender and are provided with flagella. They fertilize the eggs, and the resulting zygotes take on an active wormlike form. They wriggle through the stomach wall to its outer layer, where they become inclosed in capsules formed partly by stomach tissue. Here they grow enormously, forming little wartlike projections, which may eventually cover the outside of the mosquito's stomach. Inside its capsule each zygote divides into a number of nucleated bodies, and each of these then produces a large number of slender offspring called 'sporozoites'. Each zygote may produce as many as ten thousand sporozoites. The sporozoites migrate to the salivary glands of the mosquito and are injected into the wound when the infected mosquito bites a man. The parasites invade the red blood cells, become amoeboid, and go through the cycle described.

The disease begins with the bite by an infected mosquito. No symptoms appear for two or three weeks while the parasites are multiplying, destroying more and more red blood cells, and liberating poisonous metabolic products into the blood stream. When there are about a milliard parasites in the body, chills and fevers begin – a shivering chill followed by a burning, sweating fever. The body temperature may rise to 104°F. or even to 106°F. The attacks come regularly every 48 hours (or at other intervals, depending upon the species of parasite) because of the simultaneous liberation of all of the parasites. These periodic and weakening attacks usually last over a period of two weeks. Recovery may be followed by recurring attacks. Various drugs, including quinine, are effectively used in the treatment and prevention of malaria.

To control the disease, malaria patients must be kept in screened rooms so that they will not serve to infect mosquitoes. More

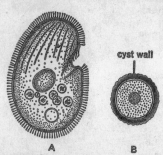

cyst wall

A   B

COLPODA. A, active animal. B, encysted animal. (B, after Kidder and Claff.)

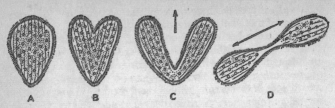

Opalina divides. A, normal animal has a large number of nuclei. B, C, D, stages in division of the cytoplasm. Division is longitudinal. Arrows show direction of swimming.

Didinium eats a paramecium. (Mostly after Calkins.)

important, the mosquitoes must be eliminated. Draining swamps, spraying ponds, or stocking ponds with fish that eat mosquito larvae are effective methods of control.

Malaria is the most important disease of man in the world as a whole. At least half the people who die from all causes are probably killed directly or indirectly by malaria. At least 250 million persons suffer from malaria every year and about 2.5 million die of it. In the United States there are perhaps a million cases of malaria a year, largely confined to the southern states; but relatively few deaths occur, because of better treatment.

The appalling social and economic consequences of this widespread disease make it a problem that requires attack not by individuals alone but by governments. It has been suggested that the ancient Greek Empire was overthrown not by conquering peoples but by the sporozoan that causes malaria.

THE CILIATED PROTOZOA

The ciliates (class CILIATA) are distinguished from other protozoa by the possession of CILIA and by the presence of TWO KINDS OF NUCLEI. These two characteristics immediately identify the paramecium, which we have already studied in some detail, as a member of this class. Like the paramecium, most ciliates collect food by means of ciliary currents and are bacteria-feeders, although a few capture and ingest other protozoa and various micro-organisms. In many ciliates the entire body is covered with cilia, as in the paramecium; but in others the cilia are greatly reduced in number and limited to particular regions of the body. The distribution of the cilia provides a convenient basis for dividing ciliates into four orders.

The MEMBRANELLES of *Stentor* are triangular plates formed by the fusion of many cilia. (Modified after Doflein.)

In the HOLOTRICHS, of which the paramecium is a member, the cilia are all short, fairly equal in length, and evenly distributed over the surface in rows or restricted to certain regions. Most of the members of this order have trichocysts, but these structures are rare in other orders. Next to the paramecium, *Colpoda* is perhaps the most common of the freshwater holotrichs. *Opalina* is a parasitic ciliate that lives in the rectum of frogs, apparently without ill effects on its host. It lacks a mouth and obtains its food by diffusion of the intestinal contents through the cell membrane. There is no contractile vacuole. *Didinium* is a holotrich that works hard for a living; it eats almost nothing but paramecia. From the centre of the front end of the animal projects a snout which is armed with structures resembling trichocysts. Didinium swims about at top speed, 'trying' to pierce everything with which it comes into contact – plants, another Didinium, or even the glass walls of an aquarium.

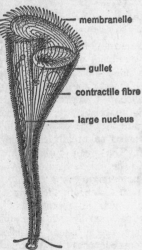

— membranelle

— gullet

— contractile fibre

— large nucleus

Stentor is a giant among protozoa; some specimens may be 1/16 inch in length. An investigator who placed a hungry *Stentor* in a thick suspension of euglenas estimated that the flagellates were taken in at a rate of about 100 per minute.

With these it has no success. When it chances to strike a paramecium, the snout penetrates and the prey is swallowed bodily.

The second group of ciliates, the *heterotrichs,* has in addition
to a covering of short cilia, a zone of large cilia around the mouth
which increases the efficiency of the feeding currents. These
large and powerful cilia, called MEMBRANELLES, are really
triangular plates formed by the fusion of a number of cilia.

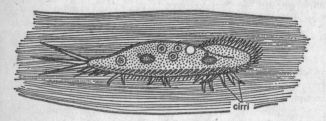

Stylonichia swims across the field of a microscope in rapid-jerks, but
it is easy to observe when it is crawling about on its cirri, as shown here.

They usually occur in a row or circlet around the mouth end of
the animal. Coincident with the development of membranelles,
these animals frequently have sessile habits, that is, they tend to
fasten themselves to the substratum. One of the most familiar
of such forms is *Stentor,* a trumpet-shaped animal which can
swim about freely but, when feeding attaches by its lower end
to a water plant or similar object, stretches out to full length,
and vibrates its circlet of membranelles so rapidly that they
look like a swiftly rotating wheel. The water current thus
produced sweeps small animals into the gullet. Stentor has a
remarkable type of large nucleus, resembling a string of beads,
and may have up to eighty small nuclei. Beneath the surface
there are lengthwise contractile fibres which shorten the animal.
Balantidium coli is the largest protozoan that lives in man and
is the only ciliate that is at all common as a human parasite.
It inhabits the large intestine, with consequences that are often
fatal and something like those resulting from the activities of
the dysentery amoeba. Balantidium occurs not only in man but
also in pigs and monkeys. Man probably becomes infected
through too close association with infected animals. Although
parasitic protozoa generally lack contractile vacuoles, Balantidium has a conspicuous vacuole. Perhaps this is related to the

fact that Balantidium unlike many parasitic forms, feeds by means of a mouth.

The differentiation of cilia seen in the heterotrichs is carried to very curious extremes in the HYPOTRICHS, which are flattened and have the cilia confined almost entirely to the lower surface. The upper surface may bear a few stiff, bristle-like cilia, but on the lower surface some of the cilia are fused in groups to form heavy CIRRI. These cirri do not beat rhythmically like ordinary cilia but are used like legs as the animal crawls about on vegetation. *Stylonichia* is one of the most common of the hypotrichs in fresh water. *Euplotes*, mentioned in chapter 3, is a hypotrich.

In another order, the PERITRICHS, the cilia are usually limited to a ring round the mouth. Some of these, like *Vorticella*, live singly, while others form branching, tree-like colonies.

Vorticella was so named because of the little vortex or whirlpool which it creates in the surrounding water by the action of the cilia about its mouth. One individual is here shown extended: the other, contracted.

The vorticellas have lost nearly all their cilia except the circlets of cilia around the broad end of the bell-shaped body. The body is firmly attached to the substratum by a slender stalk containing a spiral contractile fibre, and in the surface layer of the bell there are contractile fibres like those in Stentor. When undisturbed, the bell is poised, fully expanded, on the end of the long, straight stalk, with the cilia in rapid action. The least disturbance causes the stalk to contract like a coiled spring, while the bell also contracts, folding its edge over the circlets of cilia. The vorticellas feed chiefly on bacteria brought into the gullet by the ciliary current made by the cilia. They reproduce by lengthwise division; one of the daughters retains the stalk, while the other develops a posterior circlet of cilia and swims away, later developing a stalk and attaching itself. Any bell can also develop such a ciliary girdle, break loose from its stalk, and swim away; and vorticellas exhibit this behaviour when conditions become unsuitable. The vorticellas have one large sausage-shaped nucleus and one small nucleus. Conjugation occurs, but the two conjugating animals are of very different size.

THE SUCTORIANS

THE class SUCTORIA is a highly specialized group of protozoa which are thought to be allied to the ciliates because the young suctorians have cilia. The adults lose the cilia and are characterized by the possession of long, hollow protoplasmic extensions.

A SUCTORIAN, *Tokophrya*, feeds on a *Euplotes* many times its own size. It will take the suctorian about 15 minutes to suck its victim 'dry', and by that time the sutorian will be stretched to several times its normal size. (From A. E. Noble.)

called 'tentacles', through which they suck up the protoplasm of their prey. *Tokophrya*, a typical suctorian, is attached to the substratum by a long non-contractile stalk and bears numerous tentacles which are held rigidly extended. When a ciliate happens to come in contact with the tentacles, it is seized. The tentacles are enlarged at their ends into sucking funnels, which in some way puncture or dissolve the outer protective covering of the prey and then suck up the protoplasm.

IN this chapter we have seen that the limitations of differentiation within a small protoplasmic mass have not prevented the development of an enormously varied and extremely successful group of animals.

The phylum PROTOZOA probably shows more variation in form, in physiology, and in behaviour than the members of any other single phylum. Hence, some zoologists elevate the phylum Protozoa to the subkingdom Protozoa to distinguish them as a large and important group, with a very different plan, from all the multicellular animals.

CHAPTER 6

A Side Issue – Sponges

SPONGES, or rather the skeletons of sponges, were commonly used by the ancient Greeks for bathing, for scrubbing tables and floors, and for padding helmets and leg armour. The Romans fashioned them into paintbrushes, tied them to the end of wooden poles for use as mops, and made them serve, on occasion, as substitutes for drinking-cups. Today, sponges have an even wider variety of uses, and 'sponge-fishing' is an industry which every year produces over one thousand tons of sponges. Bath sponges grow only in warm shallow seas; but many other kinds live in the ocean depths, and some are successful in fresh waters.

A living bath sponge looks more like a slimy piece of raw liver than like the familiar sponge of the bathroom. It grows attached to the substratum like a plant, and to the casual observer shows the same kind of unresponsive behaviour. For a long time sponges were variously described as animals, plants, both animal and plant, and even as non-living substances, secreted by the many animals that take shelter in the cavities of a sponge. In fact, it was not until about a hundred years ago that the last sceptics were finally convinced of the true animal nature of sponges. The question which they had been asking and which had previously not been satisfactorily answered was: 'Since sponges do not move about and apparently do not respond rapidly to conditions about them, how can they capture food?'

This question is readily answered by adding a suspension of coloured particles to the water near a sponge, thus disclosing a great deal of unsuspected activity. A steady jet of water is seen to issue from one or more large holes at the top of the animal. Closer inspection reveals that water is at all times entering through microscopic pores that riddle the entire surface. The sponge lives like an animated filter, straining out the minute organisms contained in the stream of water that passes constantly through its body. From the possession of the millions of pores, this phylum of animals has been called the PORIFERA, or 'pore-bearers'.

UNTIL now, all the animals that we have considered were microscopic masses of protoplasm. Larger animals, like the sponges, are not merely larger masses, but their protoplasm is subdivided into microscopic units or cells. This many-celled structure is necessary for increase in the size of animals chiefly because the diffusion of oxygen and of metabolic substances is such a slow process that the interior of a large mass of protoplasm would not receive oxygen or dispose of wastes fast enough to support life. Cellular construction divides a large mass into a great number of small masses, or cells, making it possible to have spaces between them and thereby exposing an aggregate surface area many times that of the undivided bulk. This increases the surface through which substances can diffuse in and out of the protoplasm. Also, the distance that they must travel by diffusion is reduced because substances can be

A COLONY OF SIMPLE SPONGES
(*Leucosolenia*), natural size.

brought to nearly every cell in water currents. A glance at the diagram of a simple sponge shows that much more surface is exposed by this cellular construction than if the same amount of protoplasm were in a simple, solid mass.

Increase in size depends upon cellular construction for another reason. Since living protoplasm has a consistency much like that of raw egg white, a large amount of ordinary protoplasm

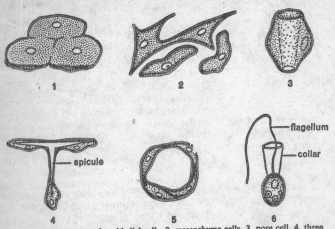

TYPES OF SPONGE CELLS. 1, epithelial cells. 2, mesenchyme cells. 3, pore cell. 4, three mesenchyme cells forming a spicule. 5, contractile cells around a pore. 6, collar cell.

could not maintain a constant shape or erect position. When the protoplasm is partitioned into microscopic cells, each enclosed in a membrane, it is more like beaten egg white, which has 'cells' of air and albumin, instead of protoplasm. A cellular protoplasmic mass can take on almost any form, especially when supported by skeletal structures.

In one-celled animals specialization is that of different kinds of intermingling protoplasms. Consequently, a protozoan is limited not only in size but also in degree of specialization. In contrast with this protoplasmic level of organization, the most primitive of the many-celled animals – the sponges – may be said to be constructed on a CELLULAR LEVEL OF ORGANIZATION. No one cell must carry on all of the life-activities, but different cells may become specialized for different functions. The various kinds of cells are not rugged individualists like the protozoa, but show definite socialistic tendencies. Only certain types feed; and these pass on some of the food to cells that specialize in protection, mechanical support, or reproduction. This division of labour among cells makes for greater efficiency and increases the possibility of exploiting sources of energy not available to simpler organisms. On the other hand, cells that specialize in certain jobs lose the ability to perform other functions and therefore become less able to lead an independent life.

Spicules from CALCAREOUS sponges.

The advantages of multicellularity are clear, but just how animals came to be composed of many cells is a question to which no definite answer can be given. One view is that as organisms grew larger and the amount of protoplasm increased, the nuclei divided, and then cell boundaries appeared. Another view is that the many-celled condition resulted from the failure of single protozoan-like cells to separate completely from each other after division. An example of this is seen in some of the colonial protozoa such as *Volvox* (illustrated on p. 51).

A SIMPLE sponge (like *Leucosolenia*) is a vase-shaped sac with a large EXCURRENT OPENING at the top and microscopic INCURRENT PORES perforating the sides. The sac is covered and protected on the outside by flattened COVERING CELLS, which fit together like the tiles in a mosaic.

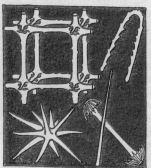

Spicules from SILICIOUS sponges.

The large internal cavity of the sponge is lined by a special kind of cell, called a COLLAR CELL because its free end is encircled by a delicate collar of protoplasm; this free end also bears a long flagellum whose base passes through the collar. The beat of the flagella of the collar cells creates the water current which passes

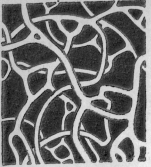

Fibres of a HORNY sponge.

through the sponge, entering by the minute pores, passing through the main cavity, and leaving by way of the large hole at the top. As the water current passes through, the collar cells capture and ingest food organisms in the same way as do certain flagellate protozoa. The flagellum undulates from base to tip, causing a stream of water to flow away from its tip. Such a stream brings particles toward the base of the cell. The collar prevents the movements of the flagellum from driving away particles as they approach the cell. Particles stick to the outside of the collar and pass down its outer surface into the cytoplasm at the base, where they are engulfed. The collar cells digest food in food vacuoles or pass it on to certain other cells that carry on digestion.

The collar cells look and behave almost exactly like the collar-flagellates (described on pp. 53-4). For this reason it is believed that sponges evolved from the same group of ancestral protozoa that also gave rise to the modern collar-flagellates.

Between the outer covering cells and the collar cells is a non-living jelly-like material containing living MESENCHYME CELLS which move about in amoeboid fashion and are more or less attached to each other by pseudopods. These are the least specialized cells and can develop into any of the more specialized types. They receive partly digested food particles from the collar cells, complete the digestion, and carry the digested food from one place to another. They probably also transport waste material to surfaces, from which it can be carried away by the outgoing current of water.

One of the chief functions of some of the mesenchyme cells is to secrete needles of calcium carbonate called SPICULES. The spicules form a skeletal framework which supports the soft cellular mass, keeps the canals from collapsing, and enables the sponge to grow to considerable size.

The spicules of sponges vary considerably in form. One sponge may have spicules of several shapes. But, as they are fairly constant for any particular species, they serve as an important criterion in identification.

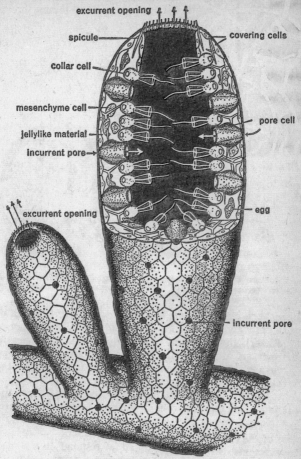

excurrent opening

spicule

covering cells

collar cell

mesenchyme cell

pore cell

jellylike material

incurrent pore

excurrent opening

egg

incurrent pore

Part of a colony of a SIMPLE SPONGE. The
upper part is cut away to show the structure.

The simple sponge, *Leucosolenia*, belongs to a group which are
known as the chalky or CALCAREOUS SPONGES because they have
spicules of calcium carbonate. The GLASS SPONGES possess siliceous
spicules. The HORNY SPONGES, which include the familiar sponges of

commerce, have a skeleton made of a horny elastic substance called 'spongin', chemically related to silk and horn.

Another type of mesenchyme cell is the PORE CELL, shaped like an old-fashioned napkin ring. In a simple sponge the pore cells lie with their outer ends opening among the covering cells and with their inner ends opening among the collar cells, so that they form the pores through which water is drawn into the sponge.

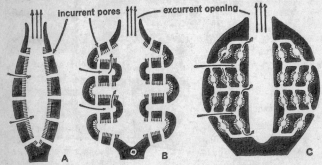

Types of sponges. A, SIMPLE SPONGE. B, MORE ADVANCED TYPE with folding of wall. C, COMPLEX SPONGE, like the bath sponge, with elaborate system of canals and flagellated chambers. (Modified after various sources.)

In some sponges there are special elongated CONTRACTILE or MUSCLE CELLS, which produce movement by becoming shorter (and thicker), thus drawing closer together the structures to which they are attached. They are arranged around openings, and, when irritating substances are present in the water, the contraction of the muscle cells narrows the openings.

Because there are no sense cells to receive stimuli from the environment and no nerve cells to transmit them to other parts of the sponge, the muscle cells must be stimulated directly in order to contract. Any sponge cell is irritable and will react if directly affected, but the animal cannot respond as a united whole. A strong touch or even an actual cut is apparently not transmitted beyond a short distance. Sponges are very insensitive to conditions about them, and, when they do react, as in decreasing the size of their openings, it is only with extreme slowness.

This lack of a specialized co-ordinating system accounts for the low grade of INDIVIDUALITY of the sponge. Many sponges

are of irregular size and shape and
consist of numerous individuals in-
extricably fused into a single large
mass or colony. An individual can
only roughly be distinguished as a
part of the sponge colony served by
one of the numerous large openings.

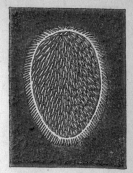

In addition to bringing a constant supply of food, the continuous current of water passing
through a sponge furnishes an
ample supply of oxygen for all the
cells and carries away carbon
dioxide and waste nitrogenous

SPONGE LARVA.

substances. Consequently, even very large sponges do not have
any special mechanisms to aid in respiration or excretion, but
simply have a larger and more efficient system of canals.

The evolution from small, simple sponges to large, complex sponges
revolves chiefly about the problem of increasing the surface in proportion to the volume. If a simple sponge like Leucosolenia were to enlarge indefinitely without any modification in structure, it would soon
reach a point at which the internal surface available for the location of
collar cells would not be large enough to bear the number of collar
cells necessary to take care of the food demands of all the other cells
composing the large bulk. In some sponges this problem has been
solved by simple folding of the body wall, which increases the surface
available for the location of the collar cells and strengthens the body
wall. In most sponges, like the bath sponges, there has been a further
increase in the folding of the walls, resulting in very intricate systems of
canals with innumerable chambers lined with flagellated cells.

Complexity serves to increase the efficiency of the sponge in another
way. A sessile animal cannot get up and leave when the food or oxygen
supplies of its environment run low. The water which leaves a sponge
has been filtered of much of its food and oxygen and is loaded with
poisonous wastes resulting from metabolism. If this water is not thrown
sufficiently far away it will enter the animal over and over again. The
evolution of the structure of sponges, then, has been in improving the
rate of flow of water through the sponge and in separating as completely as possible the outgoing from the incoming water.

Sponges may REPRODUCE SEXUALLY. Some cells of the mesenchyme enlarge greatly with reserve food and become female sex
cells or EGGS; other mesenchyme cells divide to form male sex

cells or SPERMS. In some sponges both kinds of sex cells (gametes) may arise in one individual. In others they occur in different individuals, in which case the sperms are brought into the female sponge in its water current. The fertilized egg develops into a flagellated LARVA, which finally escapes from the parent body and swims about. The term 'larva' is applied to young free-living stages in the development of animals. After swimming about for a short time, the sponge larva settles down, becomes

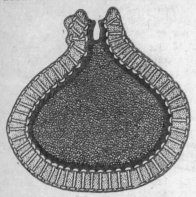

GEMMULE of a fresh-water sponge. (Modified after Evans.)

Sponge cells emerging from gemmule. (Modified after Wierzejski.)

firmly attached, and grows into a young sponge. Through their larvas the sessile sponges are able to spread geographically and to send some of their offspring far enough away from home so that they do not set up business in direct competition with their parents.

Sponges REPRODUCE ASEXUALLY by budding and branching, somewhat like plants. A simple sponge (*Leucosolenia*) sprouts horizontal branches which grow over the rocks and give rise to an extensive colony of upright vase-shaped individuals. Also buds may break off and grow into new individuals. Some sponges grow as irregular incrustations and increase their mass indefinitely over the surface of rocks, vegetation, wharf pilings or even on the backs of crabs. Thus, it is not surprising that a complete individual will grow from almost any piece which has been broken from a living sponge.

Many sponges produce asexual reproductive units known as GEM-MULES, which consist of a mass of food-filled mesenchyme cells surrounded by a heavy protective coat strengthened with spicules. The gemmule survives drying and freezing and carries the sponge over the winter season or a period of drought. Under favourable conditions the sponge cells emerge through a thin spot in the gemmule coat, aggregate in a small mass, and grow into a new sponge.

All animals – but particularly the less specialized ones – have some capacity to replace lost or injured parts, a process called REGENERATION. Some sponges are noteworthy in this respect, and

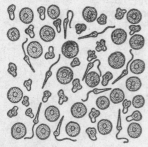

CELLS separated by pressing a living sponge through fine silk bolting-cloth.

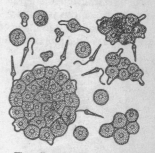

The cells move about in amoeboid fashion and aggregate in small masses which grow into small sponges. (After H. V. Wilson.)

under a certain kind of treatment their cells display a behaviour which reminds us strongly of the absurd cartoons which show a dog being ground to bits in a meat-grinder, only to emerge as small compact sausages which still retain some of the behaviour of the original dog. When certain sponges are pressed through very fine silk bolting-cloth such as is used for sifting flour in flour-mills the cells are separated from one another and come through singly or in small groups. In a dish of sea water these separated cells creep about on the bottom in amoeboid fashion. When they happen to come in contact with one another, they stick together; and after some time most of the cells are found to have united into one or more small masses. Finally these masses of aggregated cells grow up into new sponges.

THE sponge body plan is unique. No other many-celled animal uses the principal opening as an exhalant opening instead of a mouth or has the peculiar collar cells or shows so low a degree of co-ordination between the various cells. Hence, it is thought that the sponges have evolved from a group of protozoa different from the ones that gave rise to all the other many-celled animals. And the phylum PORIFERA has sometimes been set aside as a separate subkingdom of animals. There is no evidence that the sponges have ever given rise to any higher group. This does not mean that they have been a failure, for they are an abundant and widespread phylum. The 'sponge plan' is of interest to us because it illustrates the cellular level in animal structure (cell differentiation without much cell co-ordination), a stage which can no longer be found among the other many-celled animals. But in the general trend of animal evolution the sponges are little more than a side issue.

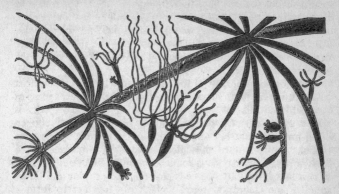

CHAPTER 7
Two Layers of Cells

THE hydra lives in ponds, lakes, and streams, attached to rocks or water plants by a sticky secretion from its disclike base. The body, with its tentacles encircling the mouth, looks like a half-inch of string with the unattached end frayed out into several strands. Because of its small size, transparency, and habit of contracting down into a little knob when disturbed, the animal is readily overlooked. Yet hydras are abundant and are the only really successful ones among the few members of their phylum that have invaded fresh water. The marine relatives of the hydras – the jelly fishes, sea anemones, and corals – are the more conspicuous members of the large and varied phylum COEL-ENTERATA. The name of the phylum is derived from 'coel', meaning 'hollow', and 'enteron', meaning 'gut', and refers to the fact that the main cavity of the body is the digestive cavity. Beginning with the coelenterates, all higher animals have a digestive cavity which connects with the outside through a mouth. For this reason, among others, it is believed that modern coelenterates, unlike sponges, evolved from the same stock that gave rise to all the higher phyla.

ONE might suppose that in the course of evolution a great many different types of cells would arise; but when animals are examined microscopically, it is found that their CELLS MAY BE CLASSIFIED INTO A FEW MAIN TYPES – about five: *epithelial, mesenchyme* or *connective, muscular, nervous,* and *reproductive.* All but nerve cells were already present in the most primitive many-celled animals, the sponges. Nerve cells are added by the coelenterates.

An association of cells of the same kind which work together to perform a common function is called a TISSUE. Thus, a mass of mesenchyme cells or some other type of connective cells is known as a 'connective tissue', a bundle of muscle cells is spoken of as 'muscular tissue', and a group of nerve cells as 'nervous tissue'. Sponges were presented as animals organized on a cellular basis, but they have some beginnings of tissue formation. For instance, the flattened cells covering the exterior and lining some of the chambers were fitted closely together to form a covering membrane. Such cells, clothing a free surface, are called 'epithelial cells'; and the tissue they form is called an EPITHELIUM.

The higher animals – man, for example – have many more different kinds of cells than a hydra, but they are all modifications of these same basic cell types. An epithelium covers the exterior surface of man, lines his mouth and digestive tract, his heart and blood vessels, and is folded in various places to form glands. Liver and thyroid cells are epithelial; blood and bone cells are mesenchyme or connective-tissue types.

The organization of cells into tissues is a distinct advance in structure, since various cell functions can be performed better by a group of cells of the same kind acting together than by separate cells. Scattered epithelial cells would be a poor protection for the surface of an animal, and epithelial cells almost always occur close together in a sheet as epithelial tissue. Similarly, a single muscle cell does not have enough strength to produce much movement, while a bundle of muscle cells contracting together can lift a heavy weight. Because their cells act together in a more co-ordinated fashion than do the cells of sponges, the coelenterates may be said

An EPITHELIUM is a group of cells covering a surface.

to have reached the TISSUE LEVEL OF ORGANIZATION.

THE hydra consists of TWO LAYERS OF CELLS. The outer layer, or ECTODERM (literally 'outside skin') is a protective epithelium, as in sponges; but it contains several other kinds of cells. The 'inner skin', or ENDODERM, lines the internal cavity and is primarily a digestive epithelium. The cells of both layers differ from ordinary epithelial cells in that their bases are drawn out into long contractile muscle fibres. The muscle fibres of the ectoderm run lengthwise.

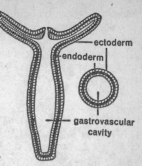

A hydra consists of TWO LAYERS OF CELLS. *Left*, a hydra in longitudinal section; *right*, in crosssection. Between the two layers is a jellylike material.

When they contract equally on all sides, the body is shortened. When they contract more on one side than on the other, the body is bent in the direction of greatest contraction. The muscle fibres in the endoderm cells run circularly; and when they contract, the body becomes narrower and longer. There is no separate muscular tissue in the hydra; the muscle fibres occur only in the bases of epithelial cells, which also perform other functions. Since they cannot be strictly classed as either epithelial or muscular, we name these cells according to function, and call those of the ectoderm the PROTECTIVE-MUSCULAR CELLS and those of the endoderm the NUTRITIVE-MUSCULAR CELLS.

Between the two layers of cells is a thin layer of jellylike material produced by both ectoderm and endoderm. Among the bases of the epithelial cells of both layers, but particularly in the ectoderm, are small MESENCHYME CELLS. The mesenchyme cells of a hydra, as in sponges, are the least specialized cells, and are capable of developing into any of the kinds of cells already mentioned, or into other kinds which will be described in connection with the hydra's activities.

In FEEDING, the hydra does not chase its prey, but remains attached to the substratum, with the almost motionless tentacles trailing in the water. When a small crustacean or worm brushes one of the tentacles in passing, the unlucky victim is suddenly riddled with a shower of poisonous, numbing threads shot out from certain of the THREAD CAPSULES, effective weapons with

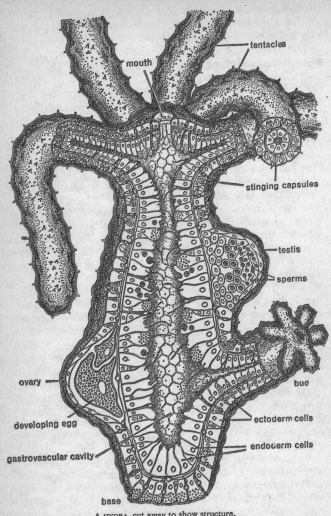

A HYDRA, cut away to show structure.

Labels: mouth, tentacles, stinging capsules, testis, sperms, bud, ovary, developing egg, gastrovascular cavity, ectoderm cells, endoderm cells, base

which the body, and particularly the tentacles, are heavily armed. There are four kinds of thread capsules, all produced within specialized mesenchyme cells, known as 'thread cells'. Each consists of a fluid-like capsule containing a long spirally-

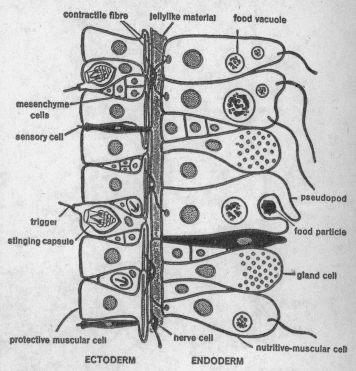

Portion of body of a hydra enlarged to show CELL TYPES. The contractile fibres in the endoderm run circularly and are shown in section as small ovals.

coiled, hollow thread. The largest and most conspicuous type (distinguished by large barbs at the base of the thread) is called a STINGING CAPSULE and lies in a cell in the ectoderm, which has a fine protoplasmic extension, or trigger, projecting from the surface. When the trigger is stimulated, a physiological change occurs, such that the pressure within the capsule is suddenly

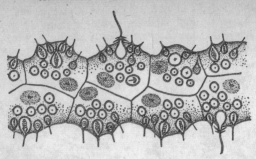

Small portion of a tentacle showing the prominent BATTERIES OF THREAD CAPSULES. Two of the large stinging capsules are discharged. Circles containing a central dot are capsules in top view. (Modified after Pauly.)

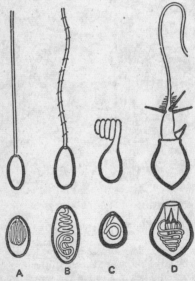

THREAD CAPSULES OF THE HYDRA. *Bottom row,* undischarged; *top row,* discharged. A, B, adhesive capsules, used to fasten the tentacles to solid objects when the hydra is looping. Probably they also help in capturing and holding prey. C, volvent type, helps in holding prey by winding about bristles of prey. D, stinging capsule, pierces body of prey and injects poison. (After various sources.)

increased and the coiled hollow thread is turned inside out. This has been compared to the way in which one everts the finger of a rubber glove by blowing into the glove. The thread is discharged with such explosive force that it pierces the body of the prey and injects a poisonous substance contained in the capsule.

Several thread capsules of the type which wind about small projections are seen clinging to the bristles of a small crustacean. (After Toppe.)

After the poison from a number of these stinging capsules has paralyzed a small animal, the tentacles are wrapped round the prey and contract, drawing the prey towards the mouth, which opens widely to receive it. The victim is swallowed by means of muscular contractions of the body wall, aided by a slimy secretion from gland cells lining the inside of the mouth region.

The mechanism by which the hydrostatic pressure within the thread capsule is suddenly raised is not definitely known. The most prevalent theory assumes an increase in pressure due to a rapid intake of water, which causes a swelling of the capsule contents. Another theory attributes the increase in pressure to the force exerted by contractile fibres surrounding the capsule.

A thread capsule (or NEMATOCYST) can be discharged only once. Used ones are discarded and are replaced by new ones, produced in specialized mesenchyme cells.

DIGESTION takes place in the interior cavity. Gland cells in the endoderm secrete enzymes, chiefly of the protein- and fat-digesting types, which reduce the digestible parts of the prey to a thick suspension containing many small fragments. This material is then engulfed by the pseudopodal activity of the nutritive-muscular cells. The process of digestion is completed within food vacuoles in these cells, for the hydra has retained, in part, the protozoan method of food ingestion and digestion. Since preliminary digestion takes place in the large digestive cavity, where enzymes poured out by many cells act together to disintegrate a food organism, a hydra can eat animals which are very large as compared with those that can be taken by a sponge. (In the sponge, as in protozoa, the prey must be of a size that can immediately be engulfed by a single cell.)

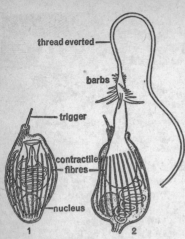

Certain thread cells show contractile fibres surrounding the capsule. 1, capsule undischarged. 2, capsule discharged. (After Will and P. Schultz.)

The indigestible remnants left in the central cavity are ELIMINATED through the mouth, which serves as both entrance and exit. The digested food is passed by diffusion from cell to cell. Currents, set up by muscular movements of the body and by the beating of long flagella on the endoderm cells, circulate the food throughout the cavity of the body and of the hollow tentacles. The cavity thus has the double function of digestion and circulation. For this reason it is called the GASTRO-VASCULAR CAVITY, which means 'stomach-circulatory' cavity.

RESPIRATION and EXCRETION take place by diffusion, as in protozoa, for the hydra is still a relatively small animal. Because of the thinness of the walls and the circulation in the gastrovascular cavity, most of the cells, inside and out, are freely exposed to the surrounding water.

When well fed and healthy, hydras REPRODUCE ASEXUALLY BY BUDDING. The buds occur about one-third the length of the body up from the base. Here both layers hump up, forming a projection which elongates and soon sprouts tentacles at its outer end. In two or three days the bud looks like a little hydra, complete with stalk, tentacles, and mouth. At its base the gastrovascular cavity of the bud is continuous with that of the parent, and in this way it receives a supply of nourishment. Shortly after this, it constricts off from the parent and takes up the serious responsibilities of an independent life.

REGENERATION would seem to be an easy matter for an animal that always has a reserve supply of unspecialized mesenchyme cells. And the hydra does, in fact, show a marked capacity for replacing tentacles or speedily repairing the more serious injuries likely to be incurred by so delicate an animal. Even if the hydra is cut into a number of pieces, most of the pieces will grow the missing parts and will become complete and independent hydras (see chap. 12).

The ability of these animals to replace lost or injured parts won for them the name of 'hydra'. An early naturalist saw in this habit a resemblance to the mythical monster, Hydra, which was finally slain by Hercules. Hydra had nine heads; and when Hercules cut one off, two grew in its place.

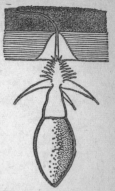

Hydras REPRODUCE SEXUALLY at certain times of the year, generally in the autumn or winter. In some species both male and female sex cells occur in the same individual, which is then known as a HERMAPHRODITE. In other species, the two sexes are always separate, and male and female individuals can be distinguished. The sex cells come from the

The thread of a stinging capsule has penetrated the hard chitinous covering of the prey and lies imbedded in the soft tissue (stippled). The clear area in the chitin represents the part that has been mechanically injured by the barbs and chemically dissolved by some substance from the capsule. (Modified after Toppe.)

mesenchyme cells in the ectoderm. In certain regions these cells suddenly start to grow rapidly, causing the body wall to bulge locally. Such bulges are known as TESTES when filled with sperm-forming cells and as OVARIES when filled with egg-forming cells. In each testis the mesenchyme cells first enlarge and then divide a number of times to form many sperms. An ovary also contains many mesenchyme cells at the start; but several of these fuse, all the nuclei but one degenerate, and the result is a single, large amoeboid egg. The amoeboid egg finally incorporates the remaining yolk-filled mesenchyme cells and becomes the spherical ripe egg, packed with food reserves that will later nourish the developing embryo.

Just how the formation of sex cells is initiated is not definitely known. That one of the factors may be low temperature is suggested by the fact that some species of hydra will produce testes and ovaries if kept in a refrigerator for two or three weeks. Sometimes abundant food appears to stimulate the development of sexual maturity.

The ripe egg breaks through the covering ectoderm and projects with its outer surface freely exposed to the water. Sperms, discharged from a testis, swim through the water and surround the egg; one enters and effects FERTILIZATION. If not fertilized within a short time after it is first exposed, the egg dies and disintegrates.

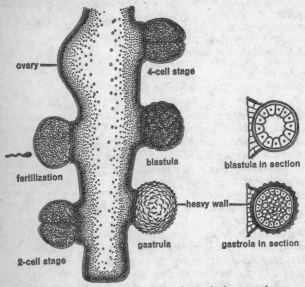

DEVELOPMENT OF A HYDRA. A hydra frequently has several eggs developing at the same time. (Based on Tannreuther and other sources.)

DEVELOPMENT begins, as in sponges and nearly all other many-celled animals, with the division of the fertilized egg or ZYGOTE into two cells. These promptly divide, forming four cells. Continued division results in a one-cell-layered hollow ball known as the BLASTULA. In the hydra the cells composing the single layer now divide, and some of the surface cells migrate inward so that cells accumulate in the interior cavity. In this two-layered stage, known as a GASTRULA, the outer layer of cells produces a heavy covering membrane or shell, and the gastrula usually drops from the parent and becomes fastened, by a sticky secretion, to the substratum. Under favourable circumstances the young hydra may hatch from the shell after a week or more. In winter the egg may lie dormant until the following spring. When development is resumed, the outer layer, or ectoderm, becomes differentiated into a protective epithelium. The inner mass of cells, or endoderm, becomes hollowed out and differentiates as a digestive epithelium. Tentacles develop, a mouth breaks through, and the young hydra hatches out by rupture of the shell.

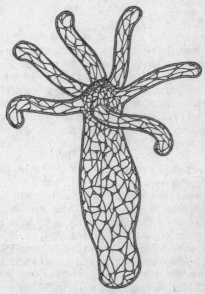

The NERVE NET of the hydra lies beneath the ectoderm. It is more concentrated around the mouth than elsewhere.

The BEHAVIOUR of the hydra is far more varied and complex than that of the uncoordinated sponge, but perhaps not much more so that that of the complex protozoa which have a system of co-ordinating fibres. The many-celled hydra has a network of nerve cells extending through the entire animal. This NERVE NET is slightly more concentrated around the mouth than elsewhere, but there is very little evidence of a specialized controlling group of nerve cells or brain which characterizes the nervous systems of more complex animals. In the nerve net of the hydra

A hydra catches ...

impulses can cross the synapses in either direction, and there are
no definite pathways for the impulses. An impulse picked up at
any place in the network can spread from one nerve cell to
another in all directions. A very strong stimulus applied to the
tip of one tentacle may cause the whole animal to contract.
This is not a very efficient kind of nervous mechanism, and the
impulses travel very much more slowly than in higher types; but
apparently it is adequate for the limited activities of the sedentary
hydra.

All the *essentials of a simple nervous mechanism* in a many-
celled animal are present in the hydra. Stimuli are received by
SENSORY CELLS which are peculiarly sensitive to touch or to
chemical substances in the water. These are slim, pointed
cells, scattered among the cells of both layers, and lying with
their pointed ends projecting to the outside if in the ectoderm,
or into the digestive cavity if in the endoderm. From the sensory
cells the impulses are picked up by NERVE CELLS which lie at the
bases of the ectoderm cells and connect with one another to form
a network extending throughout the animal. They transmit the
nervous impulses to muscle cells which contract or gland cells
which secrete.

Even non-living things can respond to stimuli. If one pushes a
rock, it may 'respond' by rolling over; but in doing so, it may
roll down a hill and break as it strikes some other rock. The
response of a hydra, or of any living organism, however, is
usually *adaptive*, that is, the response results in a favourable
adjustment of the animal to its environment. If a hydra is
touched, the sensory cells receive the stimulus. The change
started in them by the touch affects, in some way, the nerve cells
of the network, and causes them to transmit a stimulus to the

... and eats a cyclops (microscopic crustacean.)

lengthwise muscle fibres. These contract, shortening the hydra and getting it out of the way of the 'offending' object.

The nerve net not only transmits impulses but also CO-ORDINATES THE HYDRA'S ACTIVITIES. When a small animal touches one tentacle, the other tentacles will finally come to grasp the prey and will work together to cram it into the mouth. Meanwhile, impulses are going to the mouth, which opens in 'anticipation' before the food has actually touched the sensory cells around its edge. The nerve net likewise co-ordinates the muscular contractions involved in swallowing food or in forcibly removing indigestible particles. Thus it enables an animal composed of many thousands of cells to react as one integrated individual.

The importance of co-ordinated activity is emphasized by what happens when it fails. Sometimes the hydra swallows its food so rapidly that it takes in one or more of its own tentacles, and it has even been observed to swallow its own base along with the prey. Fortunately, it does not digest its own cells; and after a time the swallowed parts emerge, apparently uninjured.

The simplest method of LOCOMOTION in a hydra is a *gliding* on the base due to a creeping amoeboid movement of the basal cells. The most rapid method is a kind of *somersaulting*. The animal bends over, attaches its tentacles to the bottom by means of the adhesive thread capsules, loosens its base, swings the base over the mouth, and attaches it to the bottom; then it loosens the tentacles and repeats. In species of hydras in which the tentacles are two to five times longer than the body, the animals can move by throwing out the extended tentacles and catching hold of some object, then loosening the base, and contracting

A HYDRA SOMERSAULTING. This is its most rapid method of locomotion.

the tentacles until the body is pulled up to the object – very much like an athlete 'chinning' himself.

Hydras react to a variety of stimuli, usually by a kind of trial-and-error procedure, but almost always with a result likely to lead to the continued existence of the animal. A hydra will move away from a region of very high temperature or will migrate from the bottom of a dish to the top if the oxygen supply becomes depleted or products of decay accumulate. Many species react to light and tend to move towards the lighted side of their container, where there are usually more food organisms.

The behaviour changes with the physiological condition of the animal. A well-fed hydra usually remains attached, with the tentacles quietly extended. At intervals the body and tentacles suddenly contract and then slowly extend in a new direction. Presumably, this increases the amount of territory controlled by the animal. If a food organism does not appear after some time, the tentacles begin to wave 'restlessly', and the body contracts and expands in a new direction more frequently. If food is still not forthcoming, the hydra may move off to new hunting-grounds.

Even the food reaction varies greatly according to the state of the animal. A very well-fed hydra will not react to food when it is presented. A very 'hungry' hydra exhibits the behaviour of the feeding reaction when only the chemical stimulus – some beef juice added to the water – is present. Also, while a light touch on the tentacles may cause the food-taking reaction, a stronger touch on the tentacles or the body, as well as shaking or jarring, causes contraction of the animal in varying degrees, depending upon the strength of the stimulus. Unless the hydra has been injured by the stimulus, the body and tentacles will soon again be extended slowly, and the 'patient' life of trapping and digesting will be resumed.

OBELIA

THE nearest marine relatives of the hydras are the branching colonial coelenterates (known as hydroids) which are usually seen as delicate plantlike growths on kelps, rocks, and wharf pilings along the seacoasts. One of the commonest of these is the OBELIA, colonies of which are about an inch to several inches in height. The colony arises by budding from a single hydra-like individual. The buds fail to separate; and after repeated budding there results a treelike growth permanently fastened to some object and consisting of numerous members united by stems. Because the activities of the members are subordinate to the colony as a whole, they are sometimes called SUBINDIVIDUALS. More often a member of a colony is referred to as a POLYP, a name applied to any tubular coelenterate which bears a whorl of tentacles around the free end of the body and is attached at the other end.

The hydra is also called a 'polyp', a name which means 'many feet'. Polyps use their 'many feet', or tentacles, chiefly for feeding and only occasionally for moving about. Hence, the name is not particularly appropriate but comes to coelenterates by an indirect route. It was derived from *poulpe*, the French word for octopus, because an early French naturalist thought that coelenterate tentacles resembled the 'feet' of an octopus.

Polyps and stems are protected and held erect by a HORNY COVERING, secreted by the ectoderm, which encloses all the stems and extends around each polyp as a transparent cup, shaped like a goblet. When irritated, the polyp can withdraw into this cup; and the rapid contraction and slow expansion of the polyps are about the only movements that can be seen in an obelia colony.

OBELIA, natural size.

The stems are unable to move because of the rigidity of the covering; but at certain points the covering is arranged in rings, which allow for flexibility as the stems are swayed by water currents.

An obelia polyp is built on the same plan as a hydra, and consists of the same two cell layers, ENDODERM and ECTODERM. These are composed of cell types similar to those of the hydra.

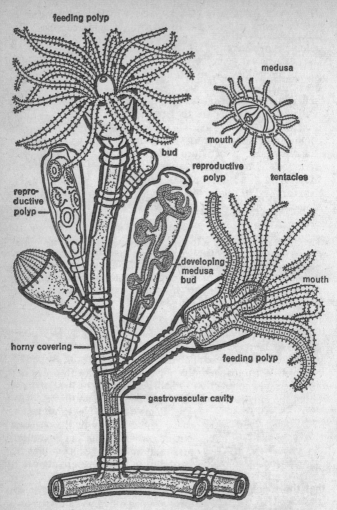

OBELIA, portion of a COLONY, and MEDUSA. One feeding polyp and one reproductive polyp have been drawn in section, showing the two layers of cells characteristic of coelenterates.

The obelia FEEDS in the same way as the hydra, capturing small prey by means of tentacles armed with stinging capsules. The tentacles are not hollow, as in the hydra, but are solid, having a central core of large endodermal cells. The polyps and stems are hollow, and the GASTROVASCULAR CAVITY of every polyp is continuous with that of every other polyp in the colony. The food is partly digested in the cavity of the polyp, and the resulting fluid is circulated about through the stems by the beating of the flagella of the digestive epithelium and by the muscular contractions of the polyps. Thus, food is distributed throughout the colony in thoroughly co-operative fashion, and digestion is completed in food vacuoles within the cells lining the gastro-vascular cavity.

ASEXUAL REPRODUCTION BY BUDDING is the usual method of increasing the number of polyps. In addition to the continuous budding on any vertical axis, rootlike horizontal stems from the base grow over the substratum and give rise to a series of upright colonies, so that the entire colony may, after a time, consist of hundreds of subindividuals.

Sexual reproduction does not occur in the polyp colony. We could search in vain throughout the year for any signs of testes or ovaries, for the obelia polyps, unlike hydras, never have any sex cells.

If we examine older obelia colonies carefully, we see that the polyps are not all alike. Those we have already described have tentacles with which they catch prey, and may be called the FEEDING POLYPS. In some of the angles where the feeding polyps branch from a stem, there occur REPRODUCTIVE POLYPS. These have lost their tentacles and the capacity to feed, and are nourished through the activities of the feeding members of the colony. They are specialized for ASEXUAL REPRODUCTION of a special type. Each is enclosed by a transparent horny vase-shaped covering and consists of a stalk on which are borne little saucer-like buds, the largest and most completely developed near the top, the smallest and least developed near the base. If live obelia colonies are kept in a dish of sea water, it is easy to observe that the topmost 'saucer' escapes through an opening at the upper end of the vase-shaped covering and swims about as a tiny animal called a MEDUSA, a name applied to any free-swimming 'jellyfish' type of coelenterate. (The name is derived from a fancied resemblance of the waving tentacles of jellyfishes to the snaky tresses of the Gorgon Medusa.)

The MEDUSA of the obelia looks like a tiny bell-shaped piece of clear jelly. From the middle of the under surface where one expects to find the clapper of a bell hangs a tube which bears

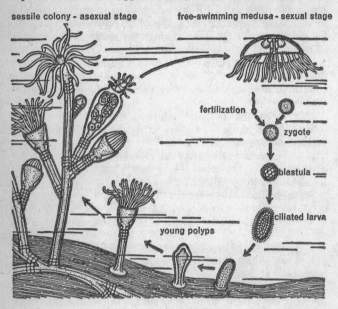

sessile colony - asexual stage   free-swimming medusa - sexual stage

fertilization

zygote

blastula

ciliated larva

young polyps

OBELIA, life-cycle.

at its free end the MOUTH. The mouth leads up this hollow tube into the GASTROVASCULAR CAVITY, which branches out into canals that carry food to all parts of the medusa. From the margin of the bell hangs a circlet of TENTACLES well armed with STINGING CAPSULES. The tiny animal swims by alternately contracting and relaxing the muscle cells of the bell. As it swims or drifts with the current, the trailing tentacles catch small organisms.

The primary function of the medusa is SEXUAL REPRODUCTION. From the under side of the bell hang four sex organs. In female medusas they are the ovaries and produce eggs; in male medusas they are testes and produce sperms. Eggs and sperms are shed into the sea water, where fertilization takes place. The zygote develops into a hollow blastula and then into a two-layered gastrula, with ectoderm and endoderm. The outer cells bear cilia, whose beating propels the little animal through the water. While the developing hydra was called an 'embryo' because it developed first attached to the parent and later within a shell, a free-swimming young stage, like that of the obelia, is called a LARVA. The larva swims about for a time, finally settles on a rock or a piece of kelp, becomes fastened at one end, develops tentacles and a mouth at the other end, and grows into a polyp which, by asexual budding, produces a new colony of sessile polyps. The larva and the free-swimming medusa both serve as a means of spreading the obelia to new localities.

Although a medusa seems very different from a polyp in general appearance, the two forms are really similar in construction. Both are composed of ectoderm and endoderm, and both consist of similar types of cells. In the medusa, ectoderm covers the entire surface of the bell and the tentacles, while the endoderm lines the various parts of the gastrovascular cavity. The chief difference from the polyp is the great thickness of the jelly between the two layers of cells. In the absence of supporting structures like the spicules of sponges or the connective tissue of higher animals, the jelly gives a firm consistency to the otherwise fragile body and adds to its buoyancy. Correlated with its greater locomotor activity, the medusa has a much more highly developed NERVE NET with a *marginal ring* of nerve cells acting as a controlling centre. There are also more specialized sensory cells. The polyp and medusa may be regarded as having adapted the same general pattern to two different ways of life – attached and free-swimming.

The occurrence of coelenterates in two forms, the medusa and the polyp, is a phenomenon termed POLYMORPHISM (meaning 'many forms'). Polymorphism is not unique among coelenterates, for many other groups, notably the social bees and ants, show structural differentiation of individuals which fits them for different roles in the life of the species. The ordinary polyps of the obelia colony capture and digest food and are capable only of asexual reproduction; they have been called feeding polyps. The stalks which bud off medusas are specialized reproductive polyps. The medusa is the sexually mature stage of the obelia and also serves to spread the species. Therefore, the obelia may be said to consist of three kinds of individuals: the *feeding polyps*, the (asexual) *reproductive polyps*, and the (sexual) *reproductive medusas*, each with its own functions. The work that in most other animals is done by every individual of the species is performed in these coelenterates by different kinds of individuals. In some colonial coelenterates, still other kinds of individuals are found, such as protective polyps which do not feed or reproduce but are heavily armed with stinging capsules.

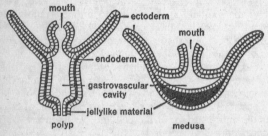

The polyp and medusa are constructed on the same general plan.

THE life-history of coelenterates may be interpreted as follows: The ancestral coelenterate was a medusa-like form which produced other medusas directly from eggs and sperms, by way of a larval stage resembling a polyp. Eventually this larval polyp-like stage became more and more important in the life-history and took on an independent existence. At first the polyps were incapable of sexual reproduction and would grow up into medusas which produced eggs and sperms. But some coelenterate polyps eventually developed the capacity for forming sex cells and dropped out the medusa stage altogether, as has happened in hydras.

The alternation of a polyp and a medusa stage has been called 'alternation of generations' or 'metagenesis'. However, it seems better not to make a distinct principle of this phenomenon. The polyp colony is simply a juvenile stage, the medusa the fully adult form, just as a caterpillar is a juvenile form capable of carrying on all activities except sexual reproduction, and the butterfly is the adult sexual form. If the caterpillar could develop mature sex organs, the butterfly stage could be dropped out of the life-history. Such a process of sexual maturity of juvenile forms is actually known to occur in quite a variety of animals besides the coelenterates, even in animals quite high up in the scale of life, such as salamanders.

THE coelenterates have only two layers of cells, the ectoderm and the endoderm. However, they show several advances in structure over the sponges because they have a network of nerve cells for co-ordinating cell activities and because they are organized on the tissue level of construction. Their most characteristic structures are the thread capsules, which are produced in no other phylum of animals. The variations among coelenterates result chiefly from a shifting emphasis in the evolution of the several groups – first upon the medusa stage, and then upon the polyp stage of the life-cycle. Some, like the obelia, have polyp and medusa almost equally developed; in others the medusa stage is more or less degenerate or wholly suppressed, as is believed to be the case with the hydra; and in others, large and well-developed medusas predominate while the fixed polyp stage is greatly reduced or has disappeared. Some of these variations are described in the next chapter.

CHAPTER 8

Polyps and Medusas

THOSE who have never seen the ocean can hardly be impressed with the coelenterates as a phylum. They have seen only the diminutive hydra and probably not at all the relatively rare fresh-water medusas. When one first walks down to a rocky ocean shore and looks in the shallow pools left by the outgoing tide, one is amazed at the display of anemones – huge, brightly-coloured polyps that look more like flowers than animals; from the algae and from rock surfaces hang delicate fringes of white, pink, or violet hydroids, like the obelia; in the open water just off shore large jellyfish drift by. But it is in warm shallow seas that the coelenterates really come into their own. There, towering growths of colonial polyps and massive banks of reef-building corals occupy almost every available square foot of the bottom, replacing the plants and dominating the lives of the other invertebrates and even of the fishes, as the trees in a forest dominate the other plants and the animals.

Coelenterates occur in two forms: polyps and medusas. The obelia described in the last chapter, shows both forms, but is more conspicuous in the polyp phase. Yet it was suggested that the primitive coelenterate was probably a kind of medusa and that the polyp was at first only a young transitory stage in the development of the medusa. Now we come to *Gonionemus*, a jellyfish that is thought to be close to the ancestral coelenterate. It has a well-developed medusa and a simple inconspicuous polyp.

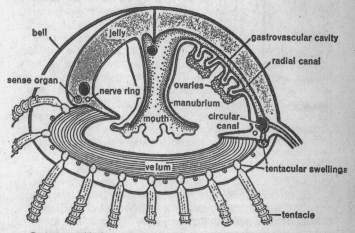

Gonionemus, with about a third of the bell removed to show internal structures.

The gelatinous bowl-shaped bell of Gonionemus has a convex outer surface and a concave under surface. From the centre of the concave surface hangs a tube, the MANUBRIUM, with the MOUTH at its tip. The other end of the manubrium leads into four RADIAL CANALS that traverse the jelly to the margin of the bell. There they join a CIRCULAR CANAL which runs around the margin and connects with the cavities of the hollow tentacles. This continuous cavity through manubrium, radial canals, circular canal, and tentacles is the GASTROVASCULAR CAVITY; it distributes partly digested food to all regions of the body.

The medusa SWIMS slowly by rhythmic pulsations of the bell. Under the ectoderm are specialized muscle fibres which do not have a protective function and serve only for contraction. The

Gonionemus.

muscular part of each individual muscle cell is very much elongated; and the epithelial part, which in hydra was so prominent, is inconspicuous here. From the margin of the bell, a muscular shelf, the VELUM, projects inward. The velum contracts strongly; and this, together with the contraction of muscle fibres in the bell, forces water out from the concavity of the bell and drives the animal in the direction opposite to that in which the water is expelled.

Gonionemus FEEDS while actively swimming about or by a kind of 'fishing' technique. The medusa swims upward, turns over on reaching the surface of the water, and then floats slowly downward with the bell inverted and the tentacles extended horizontally in a wide snare, from which passing worms, shrimps, or small fish seldom escape. When at rest, the medusa attaches to the bottom or to vegetation by the ADHESIVE PADS near the tips of the tentacles.

The free-swimming habit of Gonionemus requires greater activity and a more elaborate nervous mechanism than that of sedentary polyps like the hydra and the obelia. The NERVE NETWORK which runs beneath the ectoderm of the bell surface is concentrated around the margin of the bell into a NERVE RING, which controls and co-ordinates the animal's behaviour. Specialized SENSE ORGANS occur around the bell margin, embedded in the jelly between the bases of the tentacles. Each sense organ is a small sac containing a hard mass. As the medusa swims about, the movements of the mass within the sac probably act as stimuli which control and direct the swimming movements. In addition, the prominent swellings at the bases of the tentacles are abundantly supplied with sensory cells. These tentacular swellings are hollow and communicate with the ring canal. They are lined with pigment, which may be related to a special light-receptive function, although the whole epithelium of the lower surface is generally sensitive to light. These swellings probably serve chiefly as a site for the formation of stinging capsules, which from there migrate out along the tentacles and take their places in the stinging batteries.

The OVARIES or TESTES occur on separate female or male individuals and appear as folded ribbons which hang from beneath the four radial canals. The eggs or sperms break through the surface ectoderm and are shed directly into the water, where fertilization occurs. The zygote divides many times to form a ciliated larva, called a PLANULA, a name applied to any ciliated coelenterate larva. The planula has an outer layer of ectoderm and an inner mass of endoderm, and thus corresponds to the gastrula stage in development. It swims about for a time, then finally settles down, loses its cilia, and develops an internal cavity. A mouth breaks through at the unattached end, tentacles push out around the mouth, and the young Gonionemus soon resembles a minute, squat hydra. The polyp feeds and buds off little larvas, which also become feeding polyps. Finally, the polyps produce medusa buds which detach and grow directly into adult medusas.

The life-history of Gonionemus resembles that of the obelia, with the emphasis shifted to the medusa. Many jellyfish closely

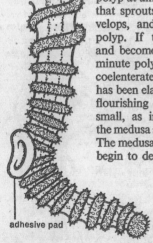

adhesive pad

End of TENTACLE of *Gonionemus* showing adhesive pad and batteries of stinging capsules.

related to Gonionemus have no attached polyp at all. One of these, *Liriope*, has a larva that sprouts its tentacles before the bell develops, and it looks like a free-swimming polyp. If this larva were to settle down and become attached, it would resemble the minute polyp stage of Gonionemus. In some coelenterates this juvenile, fixed polyp stage has been elaborated into a relatively large and flourishing colony, while the medusa is very small, as in the obelia. In other hydroids the medusa stage has been reduced still further. The medusa buds of *Hydractinia*, for example, begin to develop in the usual way but never have any medusa-like features, and fail to detach from the colony. They are degenerate saclike structures that shed eggs and sperms into the water. The final step in this direction is the complete elimination of the medusa – a condition illustrated by the hydra.

In the obelia we saw division

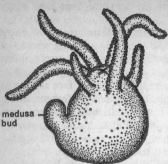

POLYP of *Gonionemus* with medusa bud.
(Based on Joseph.)

of labour not only in the life-history but also in the composition of the polyp colony. *Hydractinia* carries this POLYMORPHISM a step farther. The colony has *feeding polyps*, with mouths and long tentacles; *reproductive polyps*, which have degenerate tentacles but bear the saclike medusa buds; and *protective polyps*, which have short knoblike tentacles and cannot feed, but are richly supplied with stinging capsules that serve to protect the colony or to aid in paralysing prey.

It is among the SIPHONOPHORES, however, that we find the extremes to which colonial organization can be carried. These complex floating colonies have not only more than one kind of polyp but also more than one kind of medusa. In addition to the sexual medusas (either free or attached) they may have numerous modified medusas, called 'swimming-bells', which cannot feed or reproduce but serve only to propel the colony. There are also leaflike types that hang as protective flaps over the other members, and a gas-containing float which is thought to be a single transformed medusa that has lost the power to swim. *Physalia*, often called the 'Portuguese man-of-war', has no swimming-bells and is driven about by the action of the wind on its crested float. From the under side of the float there hang down into the water several kinds of specialized polyps, clusters of attached medusas, and tangles of long tentacles that may reach a length of 60 feet and are armed with especially large stinging capsules that can readily paralyse a large fish. The vivid blue float is a familiar and a very beautiful sight on the surface of warm seas all over the world – but it is not a very welcome one to swimmers, who know that the

Liriope, a medusa that develops directly from a zygote without a fixed polyp stage. (After Mayor.)

trailing tentacles can inflict serious and sometimes fatal injury on man.

Velella ('little sail') is also common in warm seas, and whole fleets of them may be driven into a bay by a strong wind. The colony looks like a single flat oval medusa with an erect sail-like projection from the upper surface. Around the margin is a single row of stinging protective polyps. The under surface is covered by numerous reproductive polyps which bud off free-swimming medusas; in the very centre hangs a single, large feeding polyp.

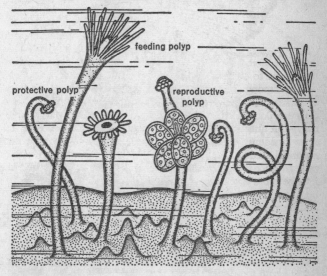

Hydractinia, a polyp colony that shows polymorphism and degenerate medusa buds. (Modified after Allman.)

In many animal groups there are marked differences between male and female individuals, and in termite colonies between individuals specialized for feeding, reproduction, and defence. These are all separate organisms – discrete physiological units. In the siphonophores it is frequently difficult to tell where one polyp ends and the next begins; and they are all closely knit into one complex organism, almost like the various organs in the body of man.

THE phylum COELENTERATA is divided into three classes. All the animals mentioned thus far belong to the class HYDROZOA ('hydra-like animals'). Most hydrozoans are polyp colonies that give rise to free medusas (*Obelia*) or to degenerate attached medusas (*Hydractinia*). But there are exceptions, like the hydra and *Gonionemus*. Further, both polyps and medusas are small, delicate, and much simpler in structure than the members of the other two classes.

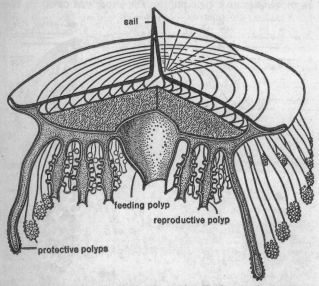

Velella, a floating colony that shows polymorphism. The front portion has been cut away to reveal the continuous gastrovascular cavity which connects the several kinds of specialized polyps. The rounded projections on the reproductive polyps are medusa buds. (Modified after Delage and Herouard.)

THE second class of coelenterates, the SCYPHOZOA ('cup animals'), includes the larger jellyfish. All are marine medusas and can be roughly distinguished from the hydrozoan jellyfish by their large size and by the absence of a velum. Moreover, the polyp stage either is lacking altogether or is very small.

Aurelia is one of the commonest of the scyphozoan jellyfish and occurs all over the world. From ocean liners one often sees large shoals of them drifting along together or swimming slowly by rhythmic contractions of the shallow, almost saucer-shaped bell. They range in size from less than 3 inches to about 12 inches across the bell. Exceptional individuals may reach a diameter of 2 feet.

At the end of a very short manubrium is a square mouth, the corners of which are drawn out into four trailing MOUTH LOBES. Each lobe has a ciliated groove. Stinging capsules in the lobes paralyse and entangle small animals, which are then swept up the grooves, through the mouth, into a spacious cavity in the

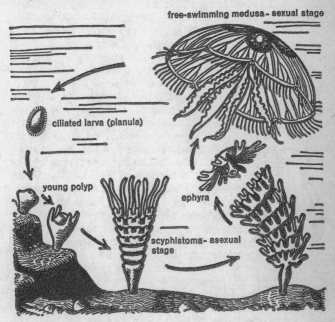

free-swimming medusa - sexual stage

ciliated larva (planula)

young polyp

ephyra

scyphistoma - asexual stage

Aurelia, life-cycle. (Based on various sources.)

centre of the bell, and from there through branched radial canals to the margin of the bell. Flagella lining the entire gastro-vascular cavity maintain a steady current of water, which brings a constant supply of food and oxygen to, and removes wastes from, the internal parts of this large animal.

The central cavity extends into four pouches, in which there are tentacle-like projections of the endoderm, called GASTRIC FILAMENTS. These are covered with stinging capsules which paralyse prey that arrives in the pouches still alive and struggling. The presence of these filaments is one of the characters that distinguish a scyphozoan from a hydrozoan medusa.

The margin of the bell bears a fringe of short and very numerous tentacles, set closely together except where they are interrupted by eight equally spaced notches. In each notch lies a SENSE ORGAN, consisting of a pigmented eyespot, sensitive to light; a hollow sac, containing hard particles whose movements set up stimuli that direct the swimming movements; and two pits, lined with cells that are thought to be sensitive to food or to other chemicals in the water.

The four horseshoe-shaped coloured bodies by which Aurelia is usually recognized are the TESTES or OVARIES, which occur (in separate individuals) on the floor of the large central part of the gastrovascular cavity. In a male medusa the sperms are discharged into the cavity and are shed to the outside through the mouth. In the female the eggs are shed into the cavity and are fertilized there by sperms which enter with the food current. The fertilized eggs go out of the mouth and lodge in the folds of the mouth lobes, where they continue to develop. The ciliated larva escapes and eventually settles down on a rock or on seaweed. There it grows into a small polyp with long tentacles. The polyp feeds and stores food, and may survive in this stage for many months, meanwhile budding off other small polyps like itself. At certain seasons, usually autumn and winter, it develops a series of horizontal constrictions which gradually deepen until the polyp resembles a pile of saucers. One by one the 'saucers' pinch off from the parent and swim away as little eight-lobed medusas, which gradually develop into adult aurelias.

This type of development in which the fixed polyp stage (hydratuba) becomes an elongated and deeply constricted polyp (scyphistoma), which successively splits off young eight-lobed medusas (ephyras), is characteristic of Scyphozoa and does not occur in the Hydrozoa.

The stinging capsules of Aurelia do not readily penetrate the human skin; but even a small *Cyanea* can raise huge weals on the arms or legs, and the monster orange and blue cyaneas of the North Atlantic are a real danger to swimmers. The largest one on record had a disc 12 feet in diameter and trailing tentacles over 100 feet long. Such huge masses of jelly are among the largest

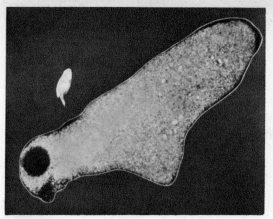

AN AMOEBA. Under indirect illumination granular protoplasm and outer membrane appear white on a dark background. The clear outer protoplasm, lacking visible granules, appears black, as does the contractile vacuole. This species is unusual in having numerous nuclei (large white dots among smaller granules). Although one-celled, amoebas are not the smallest animals. This species, a giant among amoebas, dwarfs the many-celled rotifer just above it. To the naked eye it appears as a clearly visible white speck. (Photo of living animal by P. S. Tice)

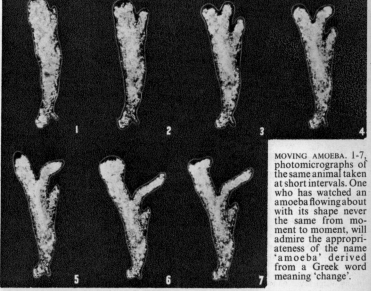

MOVING AMOEBA. 1-7, photomicrographs of the same animal taken at short intervals. One who has watched an amoeba flowing about with its shape never the same from moment to moment, will admire the appropriateness of the name 'amoeba' derived from a Greek word meaning 'change'.

(Photos of living animals by P. S. Tice)

1

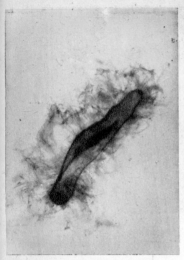

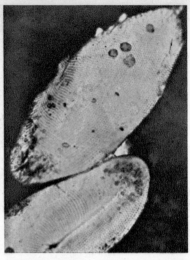

TRICHOCYSTS DISCHARGED from a paramecium irritated and later killed by a drop of ink added to water in which the animal was swimming. Said to be protective, trichocysts appear more useful in anchoring the animal while feeding. (Photo by P. S. Tice)

OUTER COVERINGS OF TWO PARAMECIA stained to show the surface markings which correspond to the positions of the cilia in the living animal. The lower covering shows the pattern of cilia around the oral groove. (Photo of stained preparation by P. S. Tice)

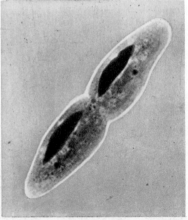

EJECTION OF SOLID WASTES through the anal pore of a paramecium. The pore has a fixed position posterior to the oral groove and is thought to be a weak spot in the outer layer of the animal. (Photo of living animal from motion picture by P. S. Tice)

PARAMECIUM DIVIDING. The large nuclei appear as two large, dark elongated bodies joined by a delicate strand. The small nucleus is already completely divided into two small, dark bodies. The cytoplasm has begun to pinch in two. (Photo of stained preparation)

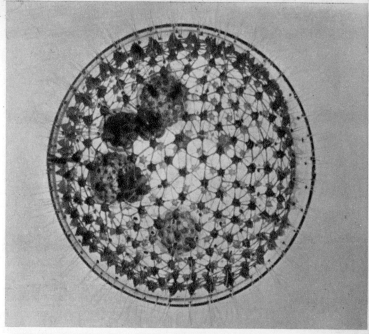

A PROTOZOAN COLONY, *Volvox,* is seen in ponds as a green sphere, about 1/25 inch in diameter. It swims by the action of flagella, co-ordinated by means of protoplasmic strands between neighbouring cells. Within the hollow colony are several daughter colonies. (Glass model. Photo courtesy American Museum of Natural History)

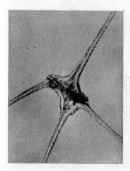

A DINOFLAGELLATE, *Ceratium,* has projections which aid in floating. Found in ponds and lakes, sometimes in tap water. (Photo of preserved specimen by P. S. Tice)

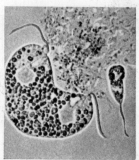

EUGLENA DIVIDING. The flagellum and anterior part of the animal have already split. A small flagellate is seen at the right. (Photo of living animals)

A PARASITIC FLAGELLATE, *Giardia,* that lives in the small intestine of man, causing diarrhoea. (Drawing, courtesy Army Med. Mus.)

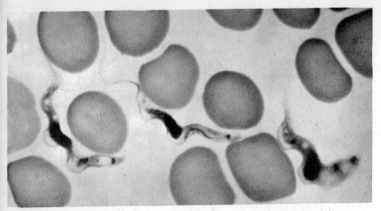

TRYPANOSOMES (*Trypanosoma gambiense*) which cause African sleeping sickness are shown here in a blood smear, among the red blood cells. The parasites are about 1/1,000 inch long. Trypanosomes probably lived originally in the intestine of blood-sucking insects, where they were constantly exposed to the vertebrate blood ingested by their insect hosts. Accidentally introduced into the blood of vertebrates by the bites of insects, the flagellates became adapted to their new environment, and finally dependent for part of their life-cycle on development in the blood stream of a vertebrate. (Photo of stained preparation. Courtesy Gen. Biol. Supply House)

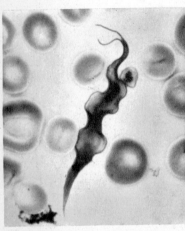

TRYPANOSOME dividing in blood of rat. The flagellum is already divided but the nucleus is still single. This flagellate (*Trypanosoma lewisi*) does not harm the rat. When a rat is infected, the parasites multiply rapidly at first, but gradually the rat's blood develops the property of inhibiting their reproduction. (Photo of stained preparation. Courtesy General Biological Supply House)

VICTIM OF AFRICAN SLEEPING SICKNESS. Much of the laziness attributed to the African natives by the early explorers was no doubt due to this disease, and slave-traders early learned not to accept as slaves Negroes having swollen glands in the neck, a symptom of trypanosome infection. (Photo made in Belgian Congo. Courtesy Army Medical Museum)

4

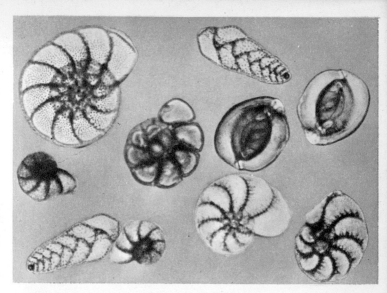

FORAMINIFER SHELLS, when highly magnified under the microscope, look like snail shells but are secreted by amoeboid protozoa. The minute pores which perforate the surface of most of the shells shown here are the openings through which the live animal extended pseudopods. The shell on the upper right has no pores, and its former occupant protruded all the pseudopods through a single opening at one end.

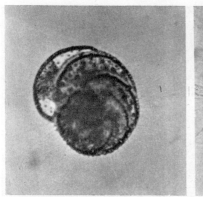

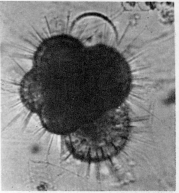

LIVING FORAMINIFER in organic debris scraped from the surface of a bit of seaweed. Long delicate pseudopods, by which the animal feeds, can be seen extending through the shell. (Pacific Grove, California, U.S.A.)

GLOBIGERINA is one of the most common foraminifers. The radiating processes are spines on the shell. The pseudopods are all withdrawn into the shell. (Photo of living animal. Pacific Grove, California, U.S.A.)

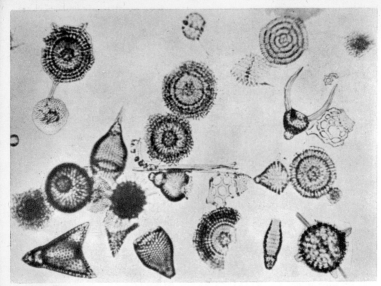

RADIOLARIAN SKELETONS are made of silica secreted by the protoplasm and usually take the form of an intricate latticework, through the openings of which are extended the pseudopods. These minute but hard skeletons constitute a part of 'Tripoli stone' which is used in abrasive powders for polishing metals.

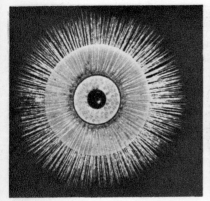

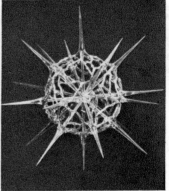

GLASS MODEL OF LIVING RADIOLARIAN. The black central body represents the nucleus. This is surrounded by a cytoplasmic capsule which, in turn, is imbedded in a frothy protoplasm. The pseudopods differ from those of foraminifers in being stiff. (Photo courtesy Amer. Mus. Nat. Hist.)

GLASS MODEL OF RADIOLARIAN SKELETON shows spherical symmetry characteristic of free-floating protozoa that cannot swim and must meet their environment on all sides. (Photo courtesy Amer. Mus. Nat. Hist.)

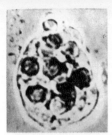

DYSENTERY AMOEBA (*Endamoeba histolytica*): *Left,* living animal containing engulfed red blood cells, from a fresh stool of an amoebic-dysentery patient. *Right,* cyst. This is the resistant form which is infectious, since it alone can survive the trip from person to person. (Photos courtesy Army Medical Museum)

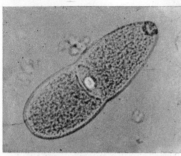

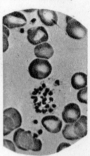

GREGARINES are sporozoan parasites of the intestine and body cavity of worms and insects. They do not live inside cells, but cling to the cells by the end shown here on the right. This one was found in the digestive glands of an acorn worm. (Photo of living animal. Bermuda)

MALARIAL PARASITE. *Left* (*Plasmodium falciparum*), female sexual form among red blood cells. *Right* (*P. vivax*), red blood cell breaking up and liberating young spores which are about to enter new red blood cells. (Photo courtesy Army Medical Museum)

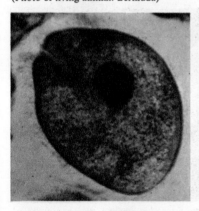
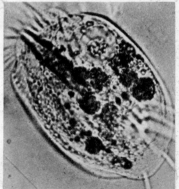

PARASITIC CILIATE (*Balantidium coli*), largest (1/300 inch) protozoan inhabiting the human intestine. It causes dysentery. The most common source of infection is the pig. (Photo courtesy Army Med. Mus.)

A HYPOTRICH does not row itself about by means of numerous fine cilia, but progresses by jerky movements of the large fused cilia on the lower surface of the animal. (Photo of living animal)

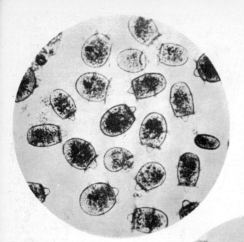

DIDINIUM, a predaceous protozoan, has a protrusible snout with which it attacks paramecia. The victims have been seen to discharge a great barrage of trichocysts without deterring the aggressive Didinium in the least. To keep healthy, Didinium must eat about eight paramecia a day, a notable feat, considering that the animal is about the same size as its prey.

STENTORS are large, trumpet-shaped ciliates which remain fixed to some object while feeding. The animals are named after a Greek herald, described in the *Iliad*, who had a very loud trumpet-like voice.

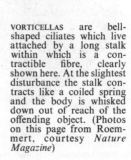

VORTICELLAS are bell-shaped ciliates which live attached by a long stalk within which is a contractible fibre, clearly shown here. At the slightest disturbance the stalk contracts like a coiled spring and the body is whisked down out of reach of the offending object. (Photos on this page from Roemmert, courtesy *Nature Magazine*)

LIVING SPONGES growing in the warm shallow waters around the Bahama Islands. The large excurrent openings are clearly visible; the small pores are microscopic. The bushy structures on the right are animals also (gorgonians, phylum Coelenterata). (Underwater photo by Johnson, Courtesy Mechanical Improvements Corp.)

COLONIAL SPONGES grow indefinitely over the surface of rocks, sending up tubular branches. (Photo of living Colony, Corona del Mar, California)

MICROCIONA, much used in regeneration experiments, is bright red and grows up to 6 inches high. (Photo courtesy American Museum of Natural History)

9

CALCAREOUS SPONGE (*Sycon*). *Above*, several individuals attached to a shell. *Below*, two enlarged; actual size ½ inch high. (Photo of preserved specimen)

HORNY SPONGE. The elephant's-ear sponge, from the Mediterranean Sea, has numerous excurrent openings on the inner wall of the cup. Its fine-grained texture and flatness (when cut up) makes it valuable to enamellers and potters. (Photo of dried skeleton)

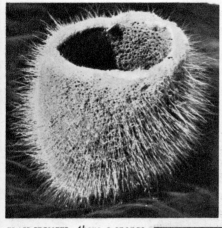

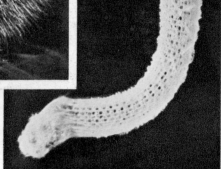

GLASS SPONGES. *Above*, a sponge with a horny skeleton and silicious spicules. *Right*, Venus' flower basket, named for the beauty of its intricate lacework of silicious spicules. This sponge grows in deep water off the Philippines. (Photos of dried skeletons)

FRESH-WATER SPONGES grow in streams, lakes, and ponds as irregular incrusting masses on sticks and stones. They grow up to the size of one's hand and are usually yellow or brown in colour but, when growing in strong sunlight, may be green from the presence of unicellular algae living in their tissues. *Spongilla*, shown here, has a horny skeleton combined with silicious spicules. (Photo courtesy American Museum Natural History)

ANOTHER FRESH-WATER SPONGE (*Ephydatia*). This specimen, collected from a Wisconsin lake, was dark grey-green in colour and only about 1 inch long. (Photo courtesy J. R. Neidhoefer and F. A. Bautsch)

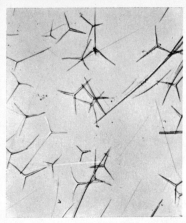

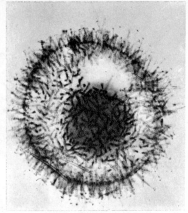

SPICULES from a calcareous sponge. The interlacing of these triradiate and single needles of calcium carbonate provides support for the soft walls of the sponge and keeps the canals from collapsing. (Photo courtesy General Biological Supply House)

THIS GEMMULE, found in a Wisconsin lake, has anchor-shaped silicious spicules, which help to protect the group of cells within. The gemmule withstands the winter; in the spring the cells emerge and grow into a new sponge. (Photo courtesy J. R. Neidhoefer and F. A. Bautsch)

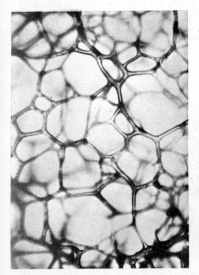

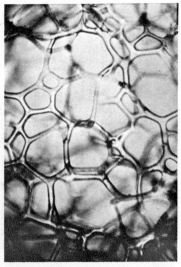

HORNY SPONGE FIBRES are elastic and form an interlacing network. *Left*, photomicrograph of a bit of *dry* sponge skeleton. *Right*, water has been added and is taken up by the fibres as well as into the spaces between fibres.

SPONGE-FISHING in the United States has its centre at Tarpon Springs, Florida, where the sponge fishermen bring in their 'catch'. The sponges are brought up by divers and after preliminary cleaning are hung on the rigging, where the rest of the protoplasm decays.

HOOKING SPONGES is a method used in shallow water. While one man rows, the other scans the bottom through a glass-bottomed bucket (which cuts off surface reflections) and pulls up the sponges with a two-pronged hook on the end of a long pole.

PREPARING SPONGES FOR THE MARKET. After being cleaned to remove the cellular debris, the sponge is pounded with a wooden mallet to break up the shells of various invertebrates that took shelter in the cavities of the living sponge. Then all irregularities are trimmed off with shears, as shown in the picture. (All photos on this page from a motion picture made by James Rhodes Co. Courtesy L. E. Diamond)

MILLIONS OF SPONGES ARE USED every year for washing automobiles, walls, floors, and ceilings and for various cleaning and polishing purposes in industry. In the home sponges are used mostly for bathing and household cleaning, and also for applying face cream and shoe polish. The tough elastic fibres of natural sponges stand hard wear much better than those of rubber or other synthetic 'sponges.' Of the thousands of different kinds of sponges in nature, only a few species of horny sponges are soft enough to be of much commercial value. Florida sponges are good enough for most cleaning purposes, but the finest ones come from the Mediterranean.

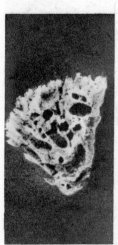

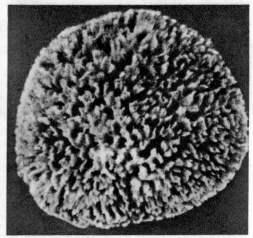

SPONGE CULTURE is being tried with some success for restocking grounds depleted by excessive sponge fishing. Live sponges are cut into small pieces (about the size of the one on the *left*), affixed to tiles, and lowered to the bottom. Under favourable conditions such pieces may grow from 2 cubic inches to 12 cubic inches in two months. The sponge on the *right* (actual size 6 inches in diameter) was 'harvested' after three years of growth in experimental 'gardens' off the coast of Florida. (From *Bulletin. U.S. Bureau of Fisheries,* Vol. 28)

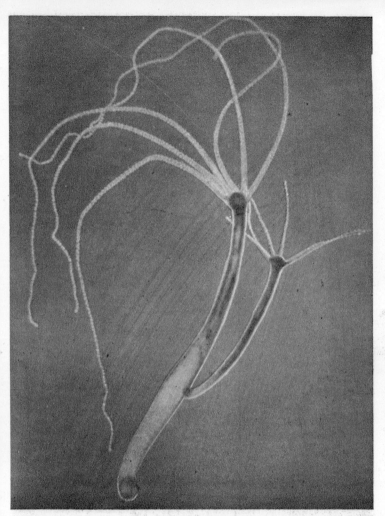

HYDRA WITH BUD. This particular species is probably *Hydra littoralis,* common in running water. The body is about $\frac{1}{2}$ inch long. When well extended, the tentacles are $1\frac{1}{2}$ times the length of the body. Other species vary in size and shape of body, number and length of tentacles (which may be shorter than, or 3 or 4 times longer than, the body), shape of stinging capsules, colour, habitat, etc. (Photo of living animal by P. S. Tice)

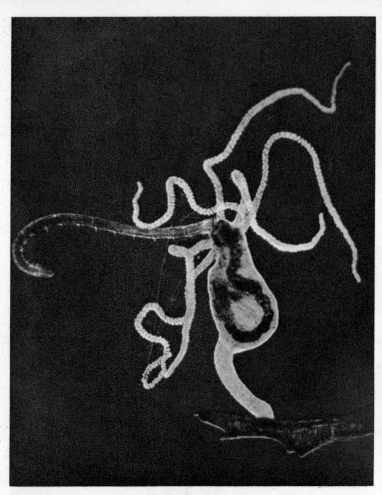

HYDRA EATING A WORM, part of which is already in the gastrovascular cavity and can be seen through the thin body wall. Besides these fresh-water aquatic relatives of the earthworm (oligochetes), the hydra eats small crustaceans, very young fish, and other small animals which come within reach of the long tentacles; the bumps on the tentacles are batteries of stinging capsules. (Photo of living animal by P. S. Tice)

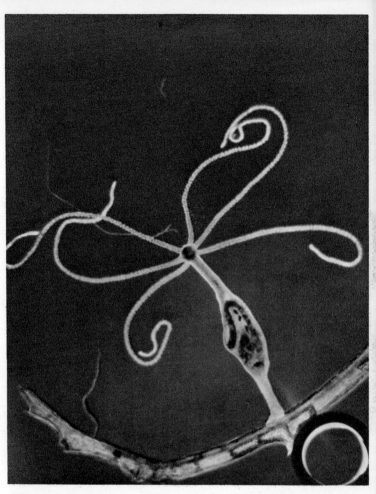

The worm safely tucked away in the gastrovascular cavity and undergoing digestion there, the hydra spreads its tentacles again and awaits a new victim. The long tentacles radiating out from a central mouth, and controlling the surrounding territory in all directions, are admirably suited to the needs of an animal that spends most of its time in one place waiting for prey to approach. (Photo of living animal by P. S. Tice)

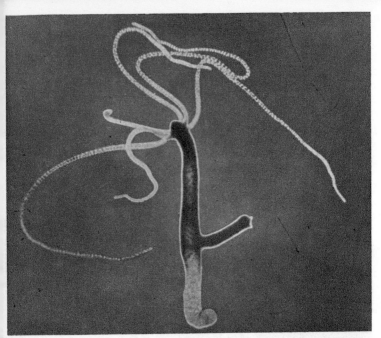

HYDRA WITH YOUNG BUD. The cavity of the bud is in direct communication with that of its parent, and the food can be seen washing back and forth. The lower third of the body contains no food and serves chiefly as a stalk which is more sharply set off from the trunk region in some hydras. (Photo of living animal by P. S. Tice)

The next day the bud is larger and has well-developed tentacles. Its behaviour is relatively independent of that of the parent. Here the parent is contracted, while the bud is extended. Photo of living animal by P. S. Tice)

18

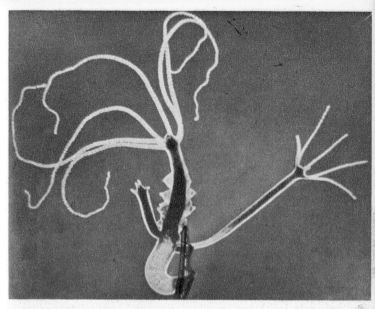

HYDRAS REPRODUCE SEXUALLY as well as asexually. This male hydra has a young bud on the left, an older bud on the right, and rows of testes along the sides of the body. (Photo of living animal by P. S. Tice)

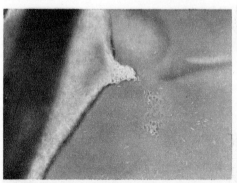

SPERMS BEING DISCHARGED FROM A TESTIS. The ectoderm is ruptured, liberating sperms which swim through the water to female hydras bearing ripe eggs. (Photo of living animal by P. S. Tice)

THREE TESTES show in this cross-section. (Photo of stained preparation, courtesy Gen. Biol. Supply House)

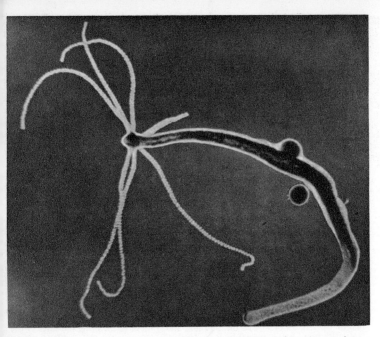

FEMALE HYDRA WITH TWO EGGS, one in the ovary and covered by the ectoderm, and one extruded. In this species the sexes are separate, but some hydras are hermaphroditic. (Photo of living animal by P. S. Tice)

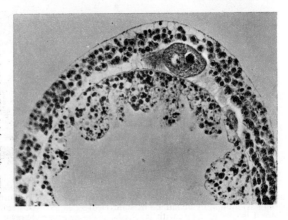

AN IMMATURE EGG lying between ectoderm and endoderm is shown in this cross-section through a female hydra. As in the stained preparation on the preceding page, only about one-half of the complete section through the animal is shown. (Photo of stained preparation, courtesy General Biological Supply House)

HYDROIDS live in shallow water attached to rocks, shells of animals, or, like the one shown here (enlarged about twice natural size), to a bit of sargassum weed, found floating in mid-ocean a few miles from Bermuda.

THE OSTRICH PLUME HYDROID (*Aglaophenia*) is commonly found on beaches where it is cast up along with seaweeds. About natural size. (Pacific Grove, California, U.S.A.)

HYDROID POLYPS WITH EXTENDED TENTACLES (highly magnified) give small animals little chance of escaping. (Photo of living colony. Bermuda)

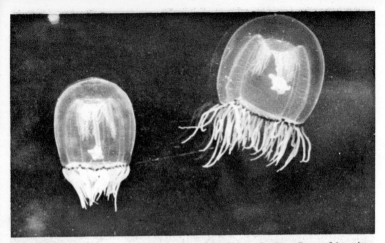

POLYORCHIS, a hydrozoan jellyfish common in bays along the West Coast of America. The mouth is at the end of the long trumpet-shaped stalk. The stringlike structures around the mouth stalk are testes or ovaries. At the base of each tentacle is a sense organ. (Photo of living animals, about natural size. Monterrey Bay, California. U.S.A.)

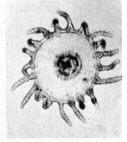

FRESH-WATER JELLY-FISH from a lake in Ohio, U.S.A. It has the habit of swimming to the surface of the water (*above*) and then coasting down (*below*) with tentacles outspread and forming a trap. (Photos of living animal about ¾ natural size taken from a film; courtesy J. A. Miller)

OBELIA JELLYFISH is only about 1/25 inch in diameter. This one was set free when a bit of an obelia colony was placed in a drop of sea water on a slide. The number of tentacles increases as it grows. *Above*, bell expanded; *below*, bell contracted. (Photos of living animal. Pacific Grove, California. U.S.A.)

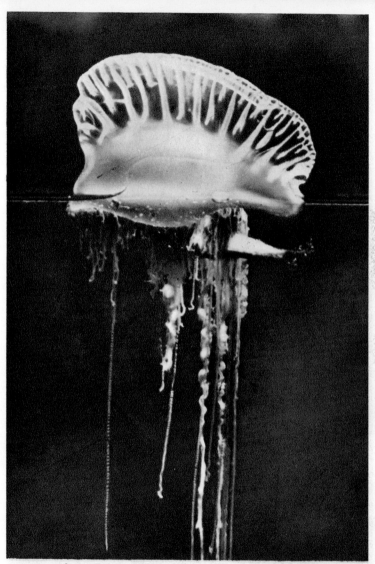

PHYSALIA, often called the 'Portuguese man-of-war', is a colonial coelenterate which moves about by the action of the wind on its gas-filled float. This one has just captured a fish, and has paralysed it with stinging capsules shot out from specialized stinging polyps. (Photo courtesy New York Zoological Society)

A JELLYFISH SWIMS BY ALTERNATELY RELAXING AND CONTRACTING THE BELL. In this photograph the bell is contracted, forcibly expelling the water from its concavity and so pushing the animal in the direction opposite to that in which the water is expelled. On the opposite, the bell is relaxed to admit water again.

Chrysaora hysocella, usually called the 'compass jellyfish' because of the marks on the upper surface of the disk, occurs in great numbers, towards the end of the summer, along the Atlantic coast of Europe. Related species are found on the coast of America. (Photos of living animal by F. Schensky, Heligoland)

CASSIOPEA, a scyphozoan jellyfish, has the lazy habit of lying in shallow water, mouth side up. This exposes the algae, which live in its mouth lobes, to the sunlight; the relationship is supposedly one of mutual benefit. The eight mouth lobes, fine branches of the gastrovascular cavity, and marginal sense organs are visible. (Photo of preserved specimens from Florida, U.S.A. Courtesy A. Novak and E. N. Miller)

STAGES IN LIFE-HISTORY OF A SCYPHOZOAN JELLYFISH. This ciliated planula, *left*, may be from almost any species of marine coelenterate. (Photo of living animal. Pacific Grove, California, U.S.A.) *Second from left*, scyphistoma showing horizontal constrictions. *Second from the right*, later stage showing young jellyfishes about to split off. *Right*, young jellyfish as seen from above. It has eight marginal lobes containing sense organs and a centrally located, four-lobed mouth. (Last three, photos of stained specimens by A. Galigher)

JELLYFISHES are often washed up on the beach in great numbers during a storm. A large jellyfish is surprisingly stiff; one can jump on such an animal without crushing it. (Photo by W. K. Fisher, Pacific Grove, California, U.S.A.)

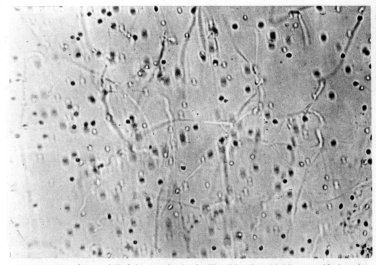

A BIT OF JELLY from a jellyfish, *Aurelia*, looks like this when highly magnified under the microscope. The dots are ameboid cells. The streaks are fibres which strengthen the jelly. The jelly consists of approximately: 95 per cent water, 4 per cent salts, and only 1 per cent organic matter. (Photomicrograph of living jelly, Bermuda)

SEA ANEMONES, so called from their resemblance to flowers, are among the most familiar coelenterates because they are easily seen in tide pools on rocky shores. The one on the right has just captured a small fish, which is held by thread capsules on the tentacles. (Photo of living animals by F. Schensky, Heligoland)

SEA ANEMONES fill every crevice in the tide pools. Animals like these have been observed to occupy the same spot for over thirty years. (Photo of living animals by W. K. Fisher, Pacific Grove, California, U.S.A.)

ANEMONE DIVIDING (*Metridium*). After a time the constriction will extend to the base, and the two anemones, asexually produced, will then separate. (Photo of living animal by D. P. Wilson, Plymouth, England)

METRIDIUM has extremely numerous delicate tentacles with which it catches minute organisms instead of larger animals like fish, on which other anemones, like the one on the lower right, feed. A large *Metridium* may attain a height of more than eight inches. The animal does not remain always in one spot but can glide on the slimy disk. One was seen to move 1½ feet in 24 hours. (Photo of living animals by F. Schensky, Heligoland)

SOLITARY CORALS. *Left,* the polyp, *Balanophyllia,* has a red stony cup about ¼ inch in diameter. The spots on the tentacles are batteries of stinging capsules. Photo of living animal. Pacific Grove, California, U.S.A.) *Right,* radial symmetry (of coelenterates) is strikingly shown by the empty cup of a small solitary coral seen from above. In the living animal the stony plates support delicate partitions which divide up the gastrovascular cavity.

A CORAL COLONY (*Astrangia*) arises by budding from a single individual. Each polyp is ⅜ inch high. *Astrangia* can grow in the cold waters off Cape Cod. (Woods Hole. Mass., U.S.A. Courtesy American Museum Natural History)

BRAIN CORAL on the ocean floor off the Bahama Islands. This colony is several feet across, but these gigantic masses of limestone may be 8 feet in diameter, or larger, and weigh several tons. (Underwater photograph by Johnson; courtesy Mechanical Improvements Corp.)

LIVE BRAIN CORAL with polyps retracted as it appears in the daytime. Most corals remain contracted during the day and expand and feed only at night. A close-up of this same coral colony at night is shown below to the left. (Bermuda)

LIVE BRAIN CORAL with partly expanded polyps, at night. The polyps do not occupy separate cups but are continuous with each other, except for separate mouths which lie at intervals along the bottom of the groove. The tentacles lining the sinuous grooves were outspread until the light was turned on to make the exposure. (Bermuda)

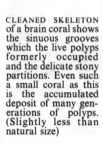

CLEANED SKELETON of a brain coral shows the sinuous grooves which the live polyps formerly occupied and the delicate stony partitions. Even such a small coral as this is the accumulated deposit of many generations of polyps. (Slightly less than natural size)

GREAT BARRIER REEF OF AUSTRALIA at low tide. The reefs, which form an underwater barrier 1,260 miles long, many miles wide, and at least 180 feet high, are a serious hazard to ships. Most of this limestone has been deposited by countless small, delicate polyps. This view is unusual in that it consists almost entirely of a single type, the stag's-horn coral (*Madrepora*). (Photo by W. Saville-Kent)

CORAL SKELETONS from the Great Barrier Reef. In life they differ even more because the polyps show a striking variety in form and in their brilliant colours. These dried and bleached skeletons are beautiful and are used as ornaments. But they give about as good an impression of the exquisite beauty of living, expanded corals as one would get of the beauty of a woman from her whitened bones. (Photo by W. Saville-Kent)

CORAL REEF exposed at low tide. Black-and-white photographs give a poor impression of the beauty of such reefs. The corals are all colours, from delicate blues and pinks to yellow-green, bright green, violet, or brown. (Photo taken on Great Barrier Reef by Mrs. C. M. Yonge)

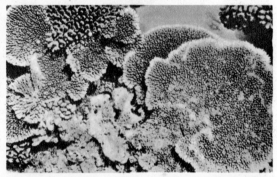

CLOSE-UP OF CORAL REEF seen above. Only the skeletons show; the delicate polyps are all retracted. They usually expand at night, spreading their tentacles to trap the small animals on which they feed. (Photo by Mrs. C. M. Yonge)

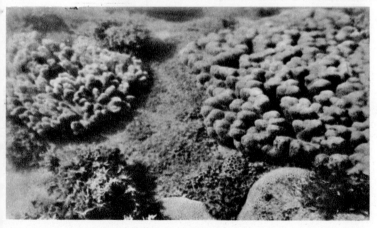

SOFT CORALS are alcyonarians, coelenterates with eight feathery tentacles. They consist of thick masses of flesh, toughened by the particles of limestone imbedded in them, and bear small delicate polyps. When the tide is out and the polyps are retracted, they look like flabby masses of leathery seaweed. (Photo taken looking down through shallow water on the Great Barrier Reef by Otho Webb)

PRECIOUS CORAL, *Corallium rubrum,* is related not to the reef corals but to the gorgonians and the pipe organ coral, as can be seen in the close-up of a single polyp, which shows the eight feathery tentacles surrounding the mouth. The anemones and true corals have simple, numerous tentacles, usually in multiples of six. The tissues of the precious coral are stiffened by limestone particles which are fused in the centre of the colony into a solid core of red limestone. It is used in making jewellery. (Photos of preserved specimens by P. S. Tice)

CORAL BLOCKS for building houses in Bermuda are cut from the hilltops with a saw. The material consists of sand, mollusk shells, and coral skeletons ground fine by wave action, deposited on the beach, and then cemented together to make the soft 'coral rock' which forms the surface layers of the island.

SEA FAN (*Eunicella*) with fully expanded polyps. As in the precious coral, the polyps have eight feathery tentacles and share the food they catch with the other members of their colony. About twice natural size. *Insert*, the whole colony, about 1/5 natural size. (Photos of living animals by D. P. Wilson, Plymouth)

GORGONIANS, often called sea whips, spring like shrubbery from among the rounded masses of corals on the ocean floor. (Photo taken off the Bahama Islands by Johnson; courtesy Mechanical Improvements Corp.)

MAKING AN UNDERWATER PHOTOGRAPH. The camera is enclosed in a watertight box held by the man in the diving helmet. (Photo by Johnson: courtesy Mechanical Improvements Corp.)

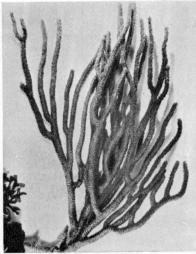

SEA WHIP brought up from 20 feet of water. The dark spots represent the positions of retracted polyps. The skeleton is not rigid as in stony corals but is quite flexible. (Photo of living colony. Bermuda)

A FULLY EXPANDED GORGONIAN occupies the centre of this picture; below it to the left is an anemone with long tentacles, and to the right are nugget and brain corals. Just above it to the left is a bouquet-shaped mass, a 'stinging coral' or millepore, the only kind of coral with capsules painful to man. Though they secrete limestone skeletons, they are related not to reef corals but to hydroids, and they have a free-swimming medusa stage. (Photo made in the Bermuda Aquarium by permission of the director, L. Mowbray)

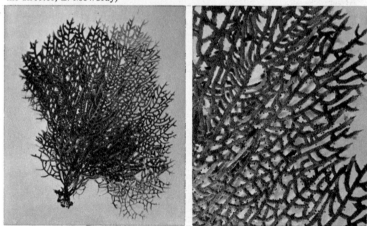

SEA FANS, whose dried purple skeletons are often used as ornaments, are gorgonians. *Left,* a small fan, about 10 inches high (they grow much taller than this). *Right,* enlarged portion of the fan to show the tiny partly retracted polyps which secrete the substance of the fan. (Photos of living colony. Bermuda)

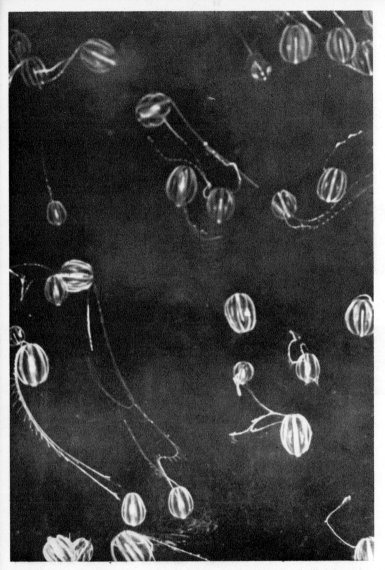

SEA GOOSEBERRIES (*Pleurobrachia*) swim by means of eight rows of ciliated combs. The two long tentacles sweep the water for small food organisms. About ⅔ natural size. (Photo of living animals by F. Schensky, Heligoland)

MNEMIOPSIS is a ctenophore or 'sea walnut' common on the East Coast of America where it is often seen in large swarms at the surface, swimming by means of eight rows of ciliated plates, nearly all of which show in the photograph. Rows of cilia, around the mouth end (toward the right in this photo), direct minute particles of food into the mouth. The animals are luminescent, and when disturbed at night, as by the passing of a boat, they light up along the rows of swimming plates. About natural size. (Photo of living animals, Woods Hole, Mass., U.S.A.)

SMALL CTENOPHORES, often called 'sea gooseberries' are frequently cast up on ocean beaches. These are *Pleurobrachia,* shown in a previous photograph with tentacles extended. (San Francisco, California, U.S.A.)

THE PLANARIA belongs to the lowest phylum of animals all the members of which possess definite heads bearing a concentration of sense organs, and have symmetry like that of man. (Photo of living animal by P. S. Tice.) *Inset* shows whole worm (*Euplanaria tigrinum*) about twice natural size. (Photo courtesy Libbie Hyman)

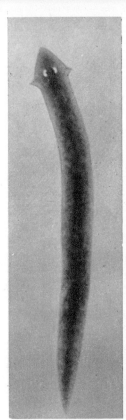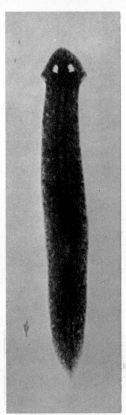

GRAFTING in animals has been studied extensively in fresh-water planarias because these animals have exceptional powers of regeneration. By producing abnormal relationships between the tissues of a single animal or by combining tissues from two different animals we learn a great deal about the development of the normal characteristics of organisms. Studying the results of interspecies grafting is particularly easy with certain planarias not only because grafts between different species take so well but because tissues of host and graft can be distinguished by their different patterns of pigmentation. *Left, Euplanaria dorotocephala,* in which the black pigment is distributed over the dorsal surface in small granules against a brownish background. The ventral surface is gray. *Middle, Euplanaria tigrina,* which has the dorsal surface finely spotted with yellowish-orange pigmented areas against a tan background. The ventral surface is white with gray pigmented areas. In the grafting experiment host and donor were placed on a tray of ice to slow down their movements. Then with a small knife a piece of the head, including the two eyes, was cut out from *E. tigrina* and placed in a hole, made just behind the eyes, in the head of *E. dorotocephala. Right,* 175 days after the operation the tissues of host and donor are completely fused. The grafted eyes lie behind those of the host. The only evidence of any effect produced by the graft on the host was the growth, in the tissues of the host, of a small projection on either side of the grafted eyes. (Photos of living animals by J. A. Miller)

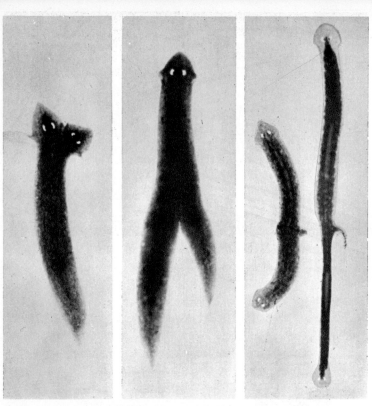

Left, an animal with two well-developed heads produced by a procedure similar to the one on the preceding page. In this case, however, the graft did not unite on all sides with the tissues of the host but has induced the formation of a complete head, having the pointed head shape, pointed sensory lobes, and pigmented pattern of the host (*Euplanaria dorotocephala*) rather than that of the donor (*E. tigrina*), which has a more rounded head with blunter lobes and a different pigment pattern. *Second from left,* a two-tailed worm obtained by grafting a head piece from *E. tigrina* into the tail region of *E. dorotocephala*. The graft grew out as a complete head with *E. tigrina* characters. It induced the formation of a pharynx in the host tail. Then the host head and pharyngeal region were removed; and the cut surface, which we might expect to regenerate a new head, produced only a pharynx and tail, supposedly because the tissue had come under the domination of the graft head. After this photograph was made, the two tails fused along their inner borders. And when this animal, with the double posterior region containing two pharynxes, later divided asexually, the new animals so produced had four eyes, two pharynxes, and a double digestive tract. *Second from right,* a graft was placed in the tail region and grew out as a head; then the posterior part of the host tore away, leaving only the graft attached to the anterior piece of the host. *Right,* the same worm, here shown greatly extended, has been fed blood to make the pharynxes stand out. The posterior pharynx which has been induced in host tissue by the graft takes a direction related to the graft head and opposite that of the original host pharynx. (Photos of living animals by J. A. Miller)

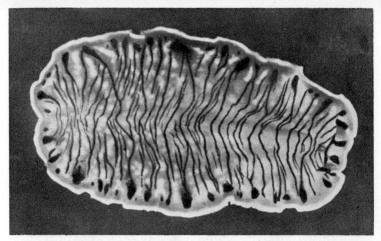

A POLYCLAD (*Pseudoceros*), pale yellow with black stripes, found on rocks or on colonies of tunicates upon which it feeds. About 1 inch long. The head, to the left, bears sensory projections. (Photo of living animal. Bermuda)

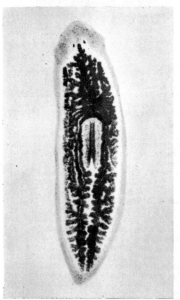

A FRESH-WATER TRICLAD (*Euplanaria*) fed a dye to make the branches of the gastrovascular cavity stand out. Actual size of worm, ¼ inch. (Photo of stained preparation)

A COMMON POLYCLAD of the West Coast of America is this extremely flattened leaflike worm. The tiny dots on the head are the numerous eyespots. Actual size of worm inch. (Photo of stained preparation)

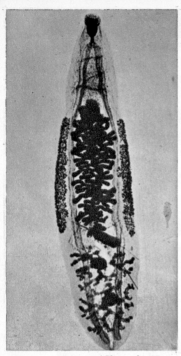

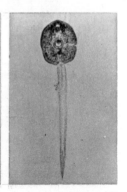

CHINESE LIVER FLUKE (*Clonorchis sinensis*) showing the two-branched gastrovascular cavity, and sex organs. Actual size, $\frac{1}{2}$ inch. (Photo of stained preparation, courtesy Gen. Biol. Supply House)

SHEEP LIVER FLUKE (*Fasciola hepatica*) stained to show the highly branched gastrovascular cavity. Actual size, $\frac{3}{4}$ inch. (Photo courtesy Gen. Biol. Supply House)

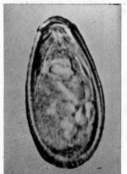

DIAGNOSIS OF FLUKE INFECTION is made by examining the excreta for eggs. *Left*, egg of a liver fluke; *right*, egg of blood fluke (*Schistosoma mansoni*). (Photo courtesy Gen. Biol. Supply House)

CERCARIA showing tail and ventral sucker. Species unknown. (Photo of stained preparation)

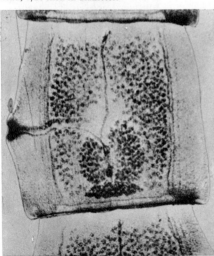

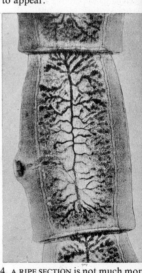

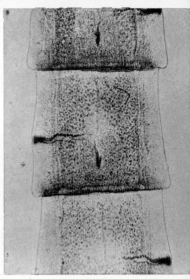

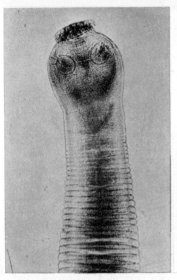

1. HEAD OF TAPEWORM (*Taenia serrata*) from a dog, showing suckers, hooks, and young sections which develop in region behind head. Actual size of head, 1/20 inch in diameter.

2. AN IMMATURE SECTION with male organs developed. The female organs are only beginning to appear.

3. A MATURE SECTION showing both sets of sex organs well developed. To identify the organs see the diagram of a typical section in the test.

4. A RIPE SECTION is not much more than a sac containing the enormously enlarged uterus filled with eggs. Actual width ⅕ inch. (Photos of stained preparations. Courtesy Gen. Biol. Supply House)

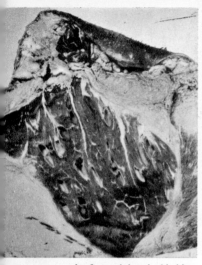

MEASLY BEEF, or beef containing the bladders of the beef tapeworm, is now rarely found on the market, thanks to government meat inspection. Meat which is cooked until it has lost its red colour is safe. (Photo courtesy Army Medical Museum)

ECHINOCOCCUS CYSTS in section of the liver of a person who accidentally swallowed the mature sections or eggs. The adult parasites live in the intestine of dogs. Actual size of cysts ½–2 inches in diameter. (Photo of specimen in Pathology Museum, University of Chicago)

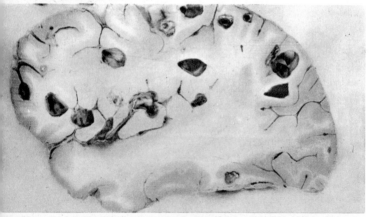

BLADDER WORMS in brain of a woman, thirty-four years old, a resident of the Chicago region, U.S.A., who was brought to the hospital with a history of epileptic fits. These became more frequent until three days before her death, when convulsions set in every half-hour. The brain (shown here in longitudinal section) contains 100–150 of these bladders. (Photo of specimen in Pathology Museum, University of Chicago)

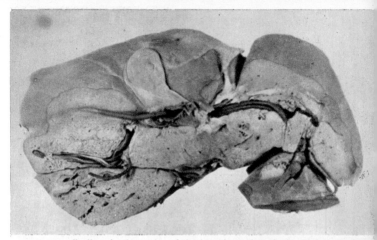

ASCARIS usually lives in the cavity of the intestine, where it does relatively little harm; but it may cause death if it wanders into the body tissues. This is a human liver cut away to show ascaris worms which have entered through the bile duct. (Photo courtesy Army Medical Museum)

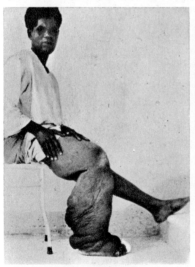

ELEPHANTIASIS is caused by certain filaria worms which live in the lymph glands and block the lymph passages. This results in diversion of lymph into the tissues and in the enormous growth of connective tissue. (Photo made in Puerto Rico by O'Connor and Hulse)

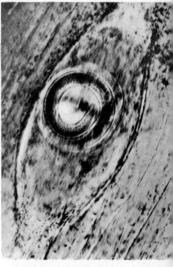

TRICHINA CYST in human muscle. The cyst does no harm, and the worm eventually dies; the damage is done by the boring of millions of these larvas before they encyst. Actual size of cyst, 1/50 inch long. (Photo of stained preparation by P. S. Tice)

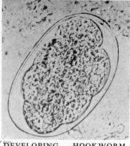

DEVELOPING HOOKWORM EMBRYO at time it leaves the body of the host is in the 4- or 8-cell stage. (Army Medical Museum)

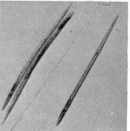

HOOKWORMS (*Necator americanus*) removed from the intestine. The worms are shown copulating; the shorter male (actual length $\frac{5}{16}$ inch) has an expansion at the posterior end by which it holds the female ($\frac{7}{16}$ inch long). (Photo courtesy Army Medical Museum)

HOOKWORM LARVAS from soil. Length 1/50 inch. This is the infective stage. (Gen. Biol. Supply House)

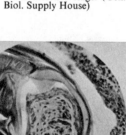

SECTION THROUGH HOOKWORM (*left*) biting wall of intestine. The American hookworm holds on by sharp cutting ridges and feeds on blood and tissue juices. *Right*, close-up of portion of same section to show the head of the worm with a bit of intestinal lining in its mouth. (Photo courtesy Army Medical Museum)

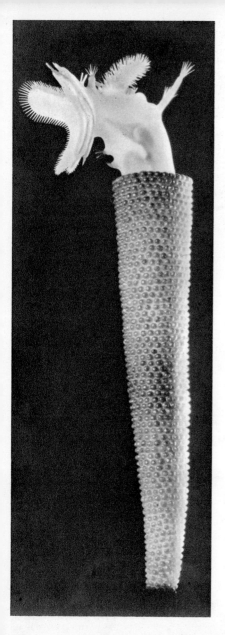

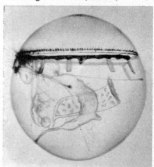

TUBE-DWELLING ROTIFER, *Floscularia,* is sessile. It builds a protective tube by cementing together minute balls of debris. Protruding from the tube is the four-lobed ciliated crown edged with cilia that create the food-bearing currents. (Model)

FLOATING ROTIFER, *Trochosphaera,* which lives at the surface of ponds and streams. The spherical body (1/50 inch in diameter) is propelled about by the band of cilia above its equator. (Glass model)

FLOATING ROTIFER, *Asplanchna,* through whose transparent body wall one can often see unborn daughters, and within them developing grandchildren! (Glass model.) (All photos on this page courtesy American Museum of Natural History)

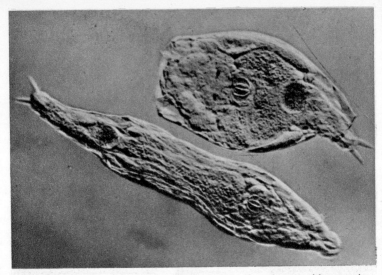

WORMLIKE ROTIFERS can be seen in almost any drop of pond water, either creeping about on vegetation or remaining temporarily attached to the substratum by means of a sticky substance secreted through the tips of the two pointed 'toes'. When fixed, the beating cilia on the lobes at the head end sweep small animals and plants into the mouth. When moving, they either swim by beating the cilia or crawl about like a leech, alternately extending and contracting the body and taking hold by the toes and then letting go. The lower animal shown here is fully extended, with the head end on the right and the toes on the left. The upper animal is contracted, with the head and foot (except the toes) telescoped into the trunk region. (Photo of living animals by P. S. Tice)

A COLONY OF ROTIFERS, *Conochilus,* appears in the upper left-hand corner of the picture. The colony swims through the water as a revolving sphere, with the members radiating from a common center at which their stalks are attached to each other. They all feed independently. (Photo of glass model courtesy American Museum Nat. Hist.)

ENCRUSTING BRYOZOANS are shown here as flat patches on the seaweed in the lower right-hand corner. (Model)

ERECT BRANCHING BRYOZOAN COLONIES are often mistaken for small delicate seaweeds. (Model)

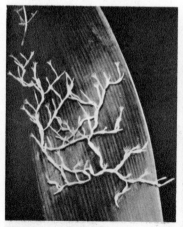

CREEPING BRYOZOAN COLONY. Each member is about 1/25 inch high. A colony makes an amusing sight with the animals popping in and out of their cases. (Photos of models. Courtesy American Museum of Nat. Hist.)

FRESH-WATER BRYOZOAN, *Pectinatella,* encrusts sticks and stones in ponds and streams. The cases of the members are jelly-like, and the colony looks like a gelatinous mass. (Photo of living colony. Illinois, U.S.A.)

BRACHIOPODS are exclusively marine animals, most of which live attached to rocks by a stalk that passes out through a hole in one valve. The two shown here are enlarged about three times. The one at the left is seen in top view; the one at the right, in side view, shows the short stalk. (Photo of living animals. Pacific Grove, California, U.S.A.)

BRACHIOPOD LAID OPEN to show the internal structure, more like that of bryozoans than like clams, which most brachiopods superficially resemble. *Laqueus californicus*, shown here about twice natural size, lives in deep water off the West Coast of America. In its natural position it lies with the ventral shell (*above*) uppermost and the valves slightly agape. The two coiled ridges supported on calcareous extensions of the dorsal valve, are bordered by delicate tentacles whose cilia maintain a circulation of water and sweep food organisms into the mouth. (Photo of preserved animal)

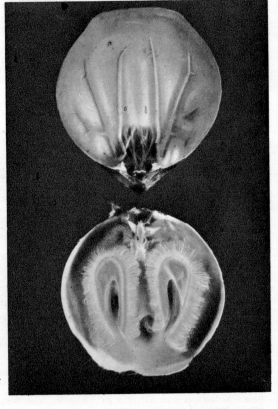

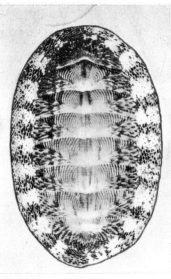
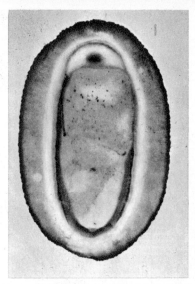

THE MOST TYPICAL MOLLUSCS are the chitons, which, though specialized in certain respects, give us some idea of the kind of animal from which snails, clams and squids have evolved. The upper surface, *left,* is protected by eight shells which overlap like the shingles on a roof. The under surface, *right,* is occupied mostly by the large, oval, fleshy foot, in front of which lies the degenerate head bearing the mouth. Surrounding the foot is the mantle, a fleshy fold which roofs over the body and secretes the shell. (Photos of living animal. Bermuda)

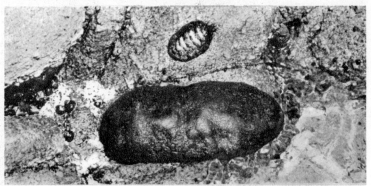

CHITONS creep about on the rocks, usually at night, rasping off fragments of algae. Most are inactive in the daytime and retreat to sheltered places, but some remain clinging to exposed rocks, where they can be seen when the tide is out. The common ones range from less than an inch long to 2–3 inches, like the smaller one shown here. The big specimen, *Cryptochiton,* about 10 inches long, is a member of the largest species of chiton. It has eight shells, but they are imbedded beneath the surface. (Photo of living animals. Pacific Grove, California, U.S.A.)

A CHITON ROLLS UP when detached from a rock – about the only defensive trick this sluggish animal has when removed from the rocky surface that normally shields the soft parts underneath. In the picture at the *left* the ventral edges of the mantle are pulled aside to show the mantle cavity, in which lie two narrow rows of gills. (Photos of living *Cryptochiton*, Pacific Grove, California, U.S.A.)

THE WHELK, *Buccinum,* is a gastropod, a member of the class of moluscs which typically have well-developed bilateral heads and coiled, asymmetrical shells and viscera. Many gastropods feed like chitons, rasping off fragments of vegetation by means of a horny toothed ribbon, the radula; but the whelk is a carnivore grasping its prey with the large muscular foot and then attacking it with a long extensible proboscis which has the radula at its tip. (The organ seen protruding from the edge of the shell is not the proboscis but the siphon, a tubular prolongation of the mantle for directing water to the gill.) It uses its proboscis to bore a hole through the hard armour of a recently dead crab or lobster. But in attacking a scallop, it simply waits until the valves are agape, and sticks the edge of its own shell between the open valves to prevent them from closing. Then it inserts the proboscis and rasps away the soft parts. (Photo of living animal by D. P. Wilson, Plymouth)

54

LAND SNAILS are gastropods that have part of the mantle cavity modified as a lung for air-breathing. They use their radulas to rasp off fragments of plants. This pair soon reduced to a few shreds the piece of lettuce on which they are shown. (Photo of living animals. Pacific Grove, California, U.S.A.)

SLUGS are land gastropods that have lost the external shell, having only a thin plate imbedded in the mantle. The slime they secrete and upon which they glide is lubricating and protective, as is demonstrated by these pictures of a slug passing unharmed over the sharp edge of a razor. (Photos by W. La Varre)

MARINE SNAILS. Periwinkles (*Littorina*) live at the water's edge and spend much of their time in air. They resist drying by retiring into the shell, closing the opening with a horny plate on the foot, and secreting a mucous seal around the shell opening. (Photo of living animals. by B. Fisher, Pacific Grove, Calif., U.S.A.)

FRESH-WATER SNAIL (*Ampullarius* from S. America) viewed from the side as it glides on a piece of glass by means of its broad, muscular foot. The eyes are on short stalks at the bases of the tentacles. This snail breathes by taking air into its lung and also by means of a gill. (Photo of living animal by P. S. Tice)

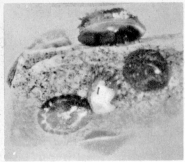

LIMPETS are gastropods that have reverted to a simple caplike shell. They are sluggish animals with habits like those of chitons. About ¾ natural size.

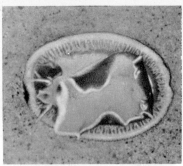

UNDERSIDE OF A LIMPET (enlarged to about double size) shows the large foot with which it creeps about on rocks. The head has two large sensory tentacles.

KEY-HOLE LIMPET, *Magathura* (5 inches long). The black mantle almost covers the shell, which has a central opening for escape of the respiratory current.

ABALONE, *Haliotis* (8 inches long), is a gastropod with a flattened spiral shell. The row of holes in the large whorl is for the outgoing respiratory current.

ABALONE SHELL HEAPS indicate the extent to which these large and delicious molluscs are eaten. The colourful pearly layer of the shell is made into jewellery.

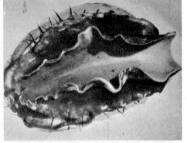

UNDERSIDE OF ABALONE shows the large muscular foot and the numerous sensory projections. (Photos on this page of living animals. Pacific Grove, California, U.S.A.)

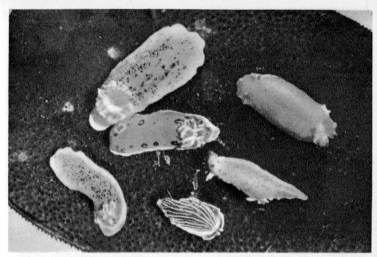

NUDIBRANCHS are sluglike marine gastropods that have small coiled shells as embryos but later lose their shells, uncoil, and develop symmetrically on the two sides – at least externally. They have lost the true molluscan gills and breathe through the feathery tuft on the upper surface near the posterior end. Most of them live among seaweeds, on which they feed; but some eat sponges. Though they look like uninteresting little lumps of flesh in this black-and-white photograph, their lovely and often bizarre shapes, soft textures, and delicate colours earn them a place among the most beautiful of all invertebrates. (Photo of living animals. Pacific Grove, California, U.S.A.)

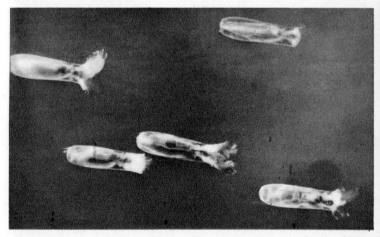

PTEROPODS are marine gastropods that swim in the open ocean by flapping finlike extensions of the foot. The uncoiled, vaselike shell is thin and transparent. Pteropods usually swim together in enormous numbers, and in northern waters some species are so abundant as to furnish food for whales. (Photo by William Beebe)

FRESH-WATER CLAMS can plough slowly through the sand or mud with the muscular foot; but they spend most of their time in one place, with the ligament up, the shells slightly agape, and the openings for water currents protruding (as in the clam at the left). Water is drawn in, strained of its load of minute food organisms, and then expelled. (Photo of living animals. Courtesy Shedd Aquarium, Chicago, U.S.A.)

HORSE-HOOF CLAM (*Hippopus*) lies with ligament down. It has a degenerate foot and never moves. A foot long, and one of the largest of clams, it is dwarfed by the giant clam, *Tridacna*, which may be 5 feet long and weigh 500 pounds. (Photo of living animal by Otho Webb, Australia)

SCALLOPS (*Pecten*) swim about erratically by clapping the two shells. Both mantle edges have a row of steely blue eyes, which show here as bright spots in the upper-most scallop. The large muscle that closes the shells is the only part of a scallop that is eaten. (Photo of living animals. Woods Hole, Massachusetts, U.S.A.)

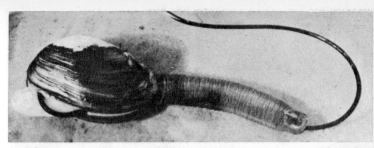

GAPER CLAM (*Schizothaerus*) is one of the largest bivalves on the west coast of America, the shell alone reaching a length of 8 inches. It lives deeply buried, with the long siphons extending to the surface of the mud. Laymen, concerned chiefly with their own digestion, find 'gapers' good eating. Biologists interested in the digestive processes of bivalves find these large clams good subjects for experimentation. This specimen (prepared by T. L. Patterson) is anaesthetized with ether; a rubber tube, with a balloon tied to its end, has been pushed into the incurrent siphon, across the mantle cavity, and through the mouth into the stomach. The tube is connected with a device for recording changes in air pressure exerted on the balloon by contractions of the clam's stomach.

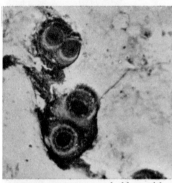

ROCK-BORING BIVALVES hold on with the foot and bore through solid rock by movements of their roughened shells. When imbedded, only the two siphons protrude. (Photo of living animals. Pacific Grove, California, U.S.A.)

WOOD-BORING BIVALVES (*Teredo*) exposed in their burrows in a piece of infested wood that has been split open. Popularly known as 'shipworms', they are not worms at all, but greatly elongated clams. The two shells, which inclose only a very small part of the anterior end of the body, have a ridged and roughened surface and are used for the boring. The animals feed on wood particles, as well as on minute organisms brought in by the respiratory current. Every year shipworms do millions of dollars' worth of damage to wooden wharf pilings and ships. (Photo of preserved specimens, twice natural size)

CLAMS, oysters, and other molluscs are our largest invertebrate source of food. In the United States squids are used by the ton for fish bait, but cephalopods are seldom eaten by man. In oriental and Mediterranean countries squids, cuttle-fishes, and octopuses are popular articles of the human diet.

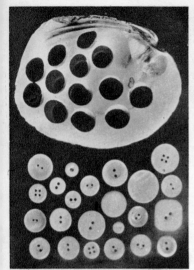

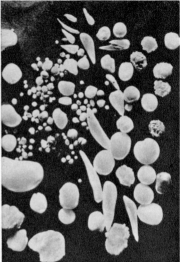

PEARL BUTTONS are cut from the shells of fresh-water clams. Thousands of tons of shells are used annually for this purpose, and in the U.S. come mostly from the Mississippi Valley. (Photo by Cornelia Clarke)

PEARLS FROM FRESH-WATER CLAMS are irregular, but occasional valuable ones are found. Pearls are protective secretions made of the same substance that lines the shell of the bivalve. (Photo by Cornelia Clarke)

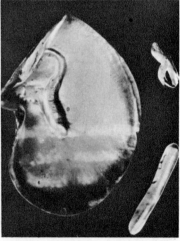

PEARLS FROM MARINE BIVALVES, like the pearl oyster, are the most valuable. About one thousand oysters are opened to find one pearl. *Left*, highly valued black pearls from the Gulf of Mexico. *Right*, mother of pearl covers a fish which became lodged between the shell and the mantle. (Photo courtesy American Museum Nat. Hist.)

THE CUTTLEFISH, *Sepia,* resembles the squid in structure and habits. The shell, a calcareous plate imbedded in the fleshy mantle, is the cuttlebone given to cage birds as a source of lime salts. The contents of the ink sac provide a rich brown pigment, sepia, used by artists. (Photo of living animal by Raoul Barba, Monte Carlo)

SQUID (*Loligo*) of East Coast of America. (Living animal. Woods Hole, Mass., U.S.A.)

SUCKER MARKS on the skin of a whale tell of an encounter between this largest of vertebrates, and the largest of invertebrates, a giant squid, which may be over 50 feet long. (Photo by L. L. Robbins)

THE NAUTILUS, in section. The animal occupies the last chamber of the coiled shell and protrudes its arms to catch crabs and other animals. It lives in deep water in the South Pacific. (2/3 natural size.) (Photo courtesy American Museum of Natural History)

LARGE SQUID. (Pacific Grove, California, U.S.A.)

THE OCTOPUS is a cephalopod with no trace of a shell. It moves by pulling itself over the rocks with its arms or by forcibly expelling water from the funnel. Its sinister reputation may be deserved by certain of the giant octopuses; but most of them, like this species (*Octopus vulgaris*, about ⅓ natural size), make for the nearest rocky crevice at the approach of a large animal like man. The sucker-bearing arms seize crabs, whose shells are then broken open by a pair of horny jaws and the radula. (Photo of living animal by D. P. Wilson, Plymouth)

EGGS OF THE OCTOPUS are each incased in a capsule, and in this species are laid in a cluster. One end of each capsule is attached to a stone or other object. The female octopus broods over the eggs. The development of the octopus (and of other cephalopods) is highly modified by the large amount of yolk in the egg and is different from that of all other molluscs. There is no free-swimming larval stage, the young octopus hatching directly from the egg capsule. In the young octopuses shown here on either side of the egg cluster, notice the prominent pigment bodies, whose contraction or expansion, under nervous control, effect rapid changes in the colour of the animal. (Photo of preserved specimens. Courtesy Gen. Biol. Supply House)

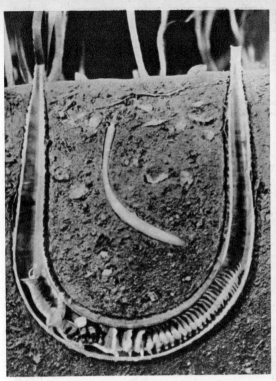

Above. TUBE-DWELLING POLYCHETE WORM, *Chaetopterus,*
lives in burrows in the mud or sand along the Atlantic
Coast of America. The U-shaped burrow, here shown in
section, is lined with a tough, parchment-like substance
secreted by the worm. The animal has a delicate body and
remains always in its well-protected burrow, feeding on
small organisms brought in by the steady current of water
that enters one end of the burrow and leaves through the
other. The current, which also brings in oxygen and carries
away wastes, is maintained by the flapping of the large,
modified parapods on the mid-region of the body. The
worm is luminescent, emitting a bluish-green light; but no
one knows how this could be of any use to an animal that
spends its life in a tube in the mud. (Photo of model
courtesy American Museum Nat. Hist.)

Right. FREE-SWIMMING POLYCHETE WORM, *Nereis,* which
also spends most of its time in a burrow in the mud or
sand but frequently leaves to build a new one, and does
swim actively at the surface during the mating season.

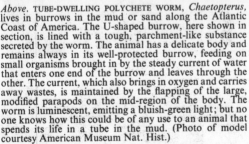

Nereis is well known to fishermen and clam-diggers of the New England coast, who
call it the clam worm because it usually occurs where clams live. Sometimes the
worms are found between the valves of an empty clam shell, and this has given rise
to the false belief that it preys on live clams. Actually, it eats much smaller animals.
(Photo by P. S. Tice)

FAN WORMS, *left* (*Bispira*), and PEACOCK WORMS, *below* (*Sabella*), are polychete worms that live in long tubes which they build in the sand, usually among rocks. Only the anterior end, a degenerate head, is extended from the tube. It bears long feathery gills which are respiratory and collect food by entangling small organisms in a layer of mucus and conveying them, by means of cilia, to the mouth. The gills are brightly coloured, usually red or purple, and a group of these worms looks like a small patch of flowers – until one approaches closely and sees them whisk the tentacles into the tube with a lightning-like speed that immediately identifies them as animals. (Photos of living animals by D. P. Wilson, Plymouth)

FEATHER WORMS, like their two relatives above, have eyespots on the gills and are very sensitive to changes in light. The shadow of one's hand passing over the extended worm, *left* will cause it to pop back into its tube, as at *right*. (Photos of living animal. Pacific Grove, California, U.S.A.)

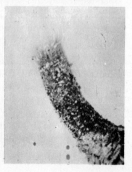

of the animals without backbones. Their long tentacles are probably responsible for some of the 'sea-serpent' stories.

THE third class of coelenterates, the class ANTHOZOA ('flower-animals'), consists of marine polyps which have no medusa stage. Anthozoa are technically distinguished from hydrozoan

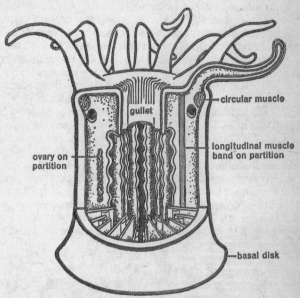

SEA ANEMONE cut away to show large gastrovascular cavity and the many partitions. The free edges of the partitions are thickened and bear gland cells.

polyps by the fact that the gastrovascular cavity is divided up by a series of vertical partitions, and the surface ectoderm turns in at the mouth to line the gullet. But superficially there is no difficulty in telling the large fleshy sea anemones or the limestone-secreting corals from most of the small fragile hydrozoan polyps.

The SEA ANEMONE has a stout muscular body, the COLUMN expanded at its upper end into an ORAL DISC having a central MOUTH surrounded by several circlets of HOLLOW TENTACLES. The basal end forms a smooth, muscular, slimy BASAL DISC on which the anemone can slide about very slowly and by which it holds to rocks so tenaciously that one is likely to tear the animal in trying to pry it loose.

From the mouth a muscular gullet hangs down into the gastrovascular cavity and is connected with the body wall by a series of vertical PARTITIONS. Between these primary partitions is a secondary set of incomplete ones, which extend only part of the way from the body wall to the gullet; and between these are still less-complete tertiary and sometimes quaternary sets. The chambers between the primary partitions are in open communication with one another below the gullet; but above the point at which they are attached to the gullet, they communicate only through one or more holes in the wall of each partition (these are shown but not labelled in the diagram of the anemone).

The partitions are double sheets of endoderm supported by a central layer of jelly, and they serve to *increase the digestive surface of the cavity,* making it possible for an anemone to digest a relatively large animal, such as a fish or a crab. The free edges of the partitions are expanded into convoluted thickenings or DIGESTIVE FILAMENTS which bear the gland cells that secrete digestive juices. Digestion is completed, as in the hydra, by pseudopodal endoderm cells lining the gastrovascular cavity.

The gullet is not cylindrical but is flattened, and at one or both ends of its long diameter is a longitudinal groove lined with cilia that are much longer than the ones lining the rest of the gullet. The cilia in these GULLET-GROOVES beat downward, drawing a current of water into the gastrovascular cavity and providing the internal parts of the anemone with a steady supply of oxygen. At the same time, the cilia lining the gullet proper beat upward, creating an outgoing current of water that takes with it carbon dioxide and other

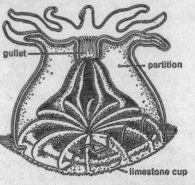

STONY CORAL POLYP. The animal has been cut away to show the gastrovascular cavity and, beneath the polyp, the beginning of the formation of the stony cup. (After Boas.)

wastes. When small animals touch the tentacles, the cilia of the gullet proper reverse their beat, and the food is swept down the gullet and into the digestive cavity.

Anemones are among the most highly specialized of the polyp types of coelenterates. They have a well-developed NERVE NET, mesenchyme cells

A colony of *Astrangia* from the Cape Cod region.

between ectoderm and endoderm, and several sets of specialized MUSCLES. A layer of circular muscles serves to narrow, and therefore to extend, the body. The longitudinal muscles are concentrated into prominent bands which run, one on each partition, from mouth to basal disc. Their contraction pulls the mouth disc, with its tentacles, completely inside. A strong circular muscle just below the mouth disc then closes over the opening, much as a pouch is drawn closed by a string. In the contracted condition anemones resist drying or mechanical injury during low tide.

Anemones sometimes REPRODUCE ASEXUALLY by pulling apart into two halves. In certain species, owing either to injury or simply to poor co-ordination, pieces of the body are left behind as the animals slide about. These fragments regenerate into tiny anemones. In SEXUAL REPRODUCTION the eggs or sperms form on the partitions of the gastrovascular cavity and are ejected through the mouth. The fertilized egg develops into a planula, which finally settles down in some rocky crevice and grows into a single anemone.

THE STONY CORALS are like small anemones but are usually COLONIAL and secrete protective LIMESTONE CUPS into which the small delicate polyps can retract. From the wall of the cup a series of radially arranged vertical plates project inward and alternate with the digestive partitions. The stony cup and its plates are, of course, outside the polyp and merely in contact with the ectoderm which secretes them.

Many kinds of small solitary cup corals grow in the temperate waters along the American coasts. Colonies of five to thirty individuals of *Astrangia* encrust shells and rocks as far north as Cape Cod. And even the cold, deep waters of the Norwegian fjords support great banks of a colonial branching coral, *Lophohelia*. But the reef-building corals grow only in tropical or subtropical waters, where the temperature never falls below 70°F.

CORAL REEFS. *Top*, a fringing reef growing around an oceanic island. *Middle*, a small barrier reef widely separated from the island. *Bottom*, an atoll. (Based on various sources.)

Three main types of coral reefs are recognized. FRINGING REEFS grow in shallow water and border the coasts closely or are separated from them at the most by a narrow stretch of water that can be waded across when the tide is out. BARRIER REEFS also parallel coasts but are separated from them by a channel deep enough to accommodate large ships. Captain Cook sailed within the Great Barrier Reef of Australia for over 600 miles without even suspecting its presence until the channel narrowed and he was wrecked on the reefs. ATOLLS are ring-shaped coral islands enclosing central lagoons, and thousands of them dot the South Pacific. Atolls are hundreds or thousands of miles from the nearest land, and their steep outer sides slope off into the depths of the ocean.

Darwin noticed that all the known coral reefs were in regions where a sinking of the land was known to have taken place or where there were evidences that it had probably occurred. He reasoned that, if an island, surrounded by a fringing reef, were to subside very slowly, so slowly that the reef could grow upwards at about the same rate, the island would grow smaller and smaller, and the fringing reef would become separated from it by a wide, deep channel, finally becoming converted into a barrier reef. If this process were to continue, the island would finally disappear entirely beneath the surface of the water, and the rising barrier reef would become a ring-shaped island, or atoll. This theory is still the most widely accepted one, though there are others almost as well in accord with known facts. In some cases an atoll may have been formed directly, without going through a fringing reef and a barrier stage, upon a submarine platform built up close to the surface by volcanic activity.

THE sea anemones and corals are zoantharians, anthozoans in which the tentacles and internal partitions are numerous and often arranged in multiples of six. There is another large group of anthozoans, the alcyonarians, in which the polyps always have EIGHT BRANCHED TENTACLES and eight internal partitions. Almost all members of this group are colonial and the body cavities of the polyps are in communication with one another through endodermal canals which penetrate the whole colony. The polyps are remarkably similar but the skeletons produced by the various colonies are strikingly diverse. The skeletons are made of minute particles, either of a horny substance or of limestone, which lie loosely in the soft tissues or grow together to form a hard, compact support, or do both in the same colony. Since any coelenterate which secretes a compact skeleton of limestone is called a 'coral' there are a number of well-known corals among alcyonarians.

The polyps of the ORGAN-PIPE CORAL live in separate vertical limestone tubes which are joined at intervals by horizontal platforms in which run the endodermal canals. The well-known PRECIOUS CORAL has a firm outer tissue, stiffened by numerous scattered limestone particles, and containing openings into which the delicate polyps can be retracted. In the centre of the colony the particles are fused into a solid, hard axial core of red limestone, which is sold as 'precious coral'.

ORGAN-PIPE CORAL. The limestone tubes are red, but the polyps are a bright green. In some places this coral is one of the important reef-builders. (Modified after Haeckel.)

The GORGONIANS (sea fans, sea plumes, and sea whips) are branching, treelike colonies built on a plan similar to that of the precious coral. Their central supporting axis is made of a flexible horny material, so that they sway gracefully with the currents and form the most conspicuous and attractive feature of the coral reefs of Florida, Bermuda, and the West Indies.

In the warm, clear waters off Bermuda it is easy to descend 15 or 20 feet in a simple diving helmet and walk about on the white coral sand among tall purple gorgonians that tower overhead and low 'bushy' ones that spring from all sides like blossoming shrubbery. Among them are massive dome-shaped heads of greenish-yellow 'brain corals', huge pink or orange anemones spreading their flower discs, and many other kinds of colourful invertebrates and fishes taking refuge in this coelenterate jungle.

CHAPTER 9

Comb Jellies

COMB jellies are transparent gelatinous animals which float in
the surface waters of the sea, mostly near shores. Being feeble
swimmers, they are carried about by currents and tides, so that
they often accumulate in vast numbers in some bay where winds
have driven them. During a storm their fragile bodies are swept
ashore by the high waves and are strewn about on the beaches.
They are, therefore, not an unfamiliar sight to people who live
on the seacoasts; and they have been given many common
names, such as 'sea gooseberries' and 'sea walnuts'. These
names describe the shape and size of two of the most common
types; but they give no suggestion of the unique character from
which has been derived the common name, 'comb jelly' and the
technical name, phylum CTENOPHORA. The ctenophores ('comb-
bearers') swim about in the water by means of eight rows of
CILIARY COMBS. Each row consists of a succession of little plates
formed of large cilia fused together at their attached ends like
the teeth of a comb. The rows radiate over the surface of the
animal from the upper pole to the lower pole, like the lines of
longitude on a globe.

The combs are lifted rapidly in the direction of the upper pole, then slowly lowered to their relaxed position. Those in each row beat one after the other from the upper toward the lower pole. All eight rows beat in unison, and the animal is slowly propelled through the water with the lower pole (mouth end) in front. The rapidly beating combs refract light and produce a constant

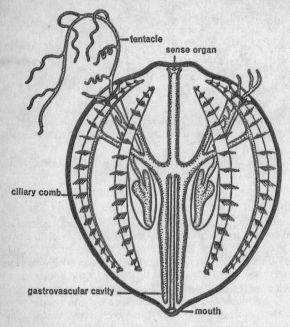

Pleurobrachia is a typical ctenophore.

play of changing colours. Comb jellies are noted for the beauty of their daytime iridescence, but this is certainly matched at night by those comb jellies that are luminescent. When the animals are disturbed as they move slowly through the dark water, they flash along the eight rows of combs.

At the upper pole of the animal is an area composed of nerve cells and sensory cells. In the centre of the area is a covered pit containing a SENSE ORGAN, which consists of a little mass of

limestone particles supported on four tufts of cilia connected
with sense cells. It is thought to act as a sort of balancing device
or 'steering-gear'. Any turning of the body causes the limestone
mass to bear more heavily upon the ciliary tuft of one side or
another. Presumably this stimulates the sensory cells, and the
stimulus is carried by nerve cells to the swimming combs, causing

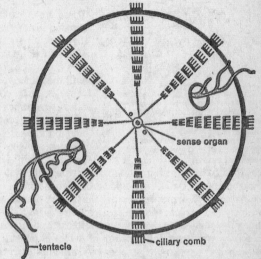

Pleurobrachia as seen from the upper pole.

them to beat faster on one side, thus righting the animal. From
this polar area a NERVE NET extends all over the body and is
concentrated into EIGHT NERVE CORDS, one under each row of
combs. This system regulates and co-ordinates the activity of
the eight ciliary rows, for, if the polar area is removed, the
combs become disorganized. And if any row is cut across, the
swimming combs below the cut get out of step with those above.

The general BODY PLAN of a comb jelly resembles that of a
coelenterate medusa. There is an epithelium of ectoderm covering the outer surface, an epithelium of endoderm lining the
gastrovascular cavity, and a thick jelly between. The jelly contains amoeboid mesenchyme cells and long, delicate muscle cells
which run from one part of the body to another.

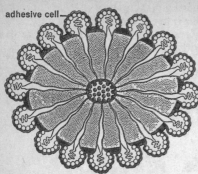

adhesive cell

Cross-section through one of the branches of a TENTACLE. The outer surface is covered by the sticky heads of the ADHESIVE CELLS. Each cell has a coiled thread (attached to the central muscular axis of the tentacle branch) which acts as a spring to prevent the cell from being wrenched off by the struggling prey. (Combined from several sources.)

The more primitive comb jellies have globular or pear-shaped bodies with a branched muscular TENTACLE on each side which can be withdrawn into a pouch. These tentacles have no stinging capsules, but the branches are covered with special ADHESIVE cells (not to be confused with the adhesive thread capsules of coelenterates), which stick to, and entangle, the prey. Such ctenophores catch small shrimps or fishes by extending their tentacles full length and then curving and looping through the water, with the sticky tentacles sweeping a wide area. Other kinds of ctenophores have the tentacles reduced to very short filaments; they feed mostly on larvas and other small organisms which are caught by the ciliated grooves and swept towards the mouth.

The mouth is situated in the centre of the lower pole and leads into a branched GASTROVASCULAR CAVITY which extends through

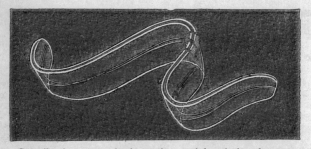

Cestus lives in warm seas but is sometimes carried north along the eastern coast of North America in the Gulf Stream. Because of the elongated shape and the beauty of the transparent body, which shimmers with blue or green in the sunlight, it is commonly called 'Venus' girdle'. (After Chun.)

the jelly, eventually giving off eight branches, one below each row of combs. Undigested food comes out through the mouth, though the gastrovascular cavity does open to the outside by two minute pores situated near the sense organ and shown (but not labelled) in the diagram of the upper pole of *Pleurobrachia*.

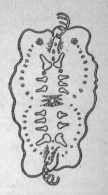

All ctenophores are HERMAPHRODITIC. Both ovaries and testes occur on the walls of each of the gastrovascular branches that run below the rows of combs. The eggs and sperms are shed to the outside through the mouth. The free-swimming larva develops directly, through rather complicated changes, into the adult.

Coeloplana, as viewed from above. This animal is common on the coasts of Japan, where it creeps about on alcyonarians. (Modified after Krempf.)

A number of bizarre forms occur among the comb jellies. *Cestus* is flattened from side to side and reaches a length of over 3 feet. It swims by muscular undulations of the ribbon-like body as well as by the beating of the elongated swimming-plates.

Coeloplana is flattened in the other direction, so that the two poles, bearing mouth and sense organ, are brought close together. It has a typical comb-jelly structure with two tentacles, but has lost its combs. By virtue of extensive development of muscle fibres it is able to creep about like a worm. Such an animal illustrates how round, free-swimming organisms can become flattened, bottom-creeping forms. As we shall see, the next phylum is characterized by the flattened, creeping type of animal.

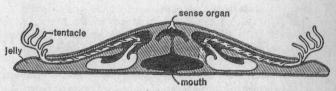

Coeloplana, as seen in diagrammatic section. The body is flattened so that the mouth and sense organs are relatively close together as compared with a more typical ctenophore like Pleurobrachia. (Modified after Komai.)

CHAPTER 10

Three Layers of Cells

PUT a piece of raw meat into a small stream or spring and after a few hours you may find it covered with hundreds of black worms that are feeding upon it. These worms, each about half an inch long, are called PLANARIAS. When not attracted into the open by food, they live inconspicuously under stones and on the vegetation.

Planarias belong to the phylum PLATYHELMINTHES ('platy', flat, 'helminthes', worms), which also includes many free-living marine species and two important groups of parasites, the FLUKES and TAPEWORMS. There are many kinds (species) of planarias, just as there are many kinds of amoebas and hydras.

The planaria differs from the hydra in that one end of the body has a definite HEAD, with eyes and other sense organs. The head is always directed forward in locomotion; and the body is clearly differentiated into front, or ANTERIOR, and rear, or POSTERIOR, ends. The planaria has an elongated flattened body; and if we watch it move, we see that one surface of the body always remains upward while the other is kept against the bottom. The upper surface is termed DORSAL (meaning back), and the lower surface, VENTRAL (meaning belly). We also notice that the various parts, such as the eyes, are symmetrically arranged on the two sides, as in ourselves.

The planaria MOVES about in a characteristic slow, gliding fashion, with the head bending from side to side as though it were 'testing' the environment. If we prod the animal, it hurries away by marked muscular waves. All these movements result from two mechanisms. The gliding is CILIARY; the other movements are MUSCULAR. The surface of the planaria consists of a protective epithelium, as in the hydra; but this is ciliated, particularly on the ventral side. Numerous gland cells, which secrete a mucous material, open on the surface of the body and pour out slime, upon which the worm moves. The cilia obtain traction on this bed of slime, and as they beat backward, they move the animal forward. Planarias do not swim freely through the water, but only in contact with a solid object or on the underside of the surface film. When a worm leaves the surface, it glides down attached to a thread of mucous. Just underneath the epithelium are layers of muscle cells. The outer layer runs in a circular direction, and the inner layer in a longitudinal direction. Muscles also run between dorsal and ventral surfaces and help

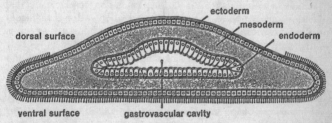

Diagrammatic cross-section of a planaria showing the general body plan.

to make possible all sorts of agile bending and twisting movements. The muscles are not part of the epithelial cells, as in the hydra, but are independent muscle cells specialized for contraction. Also, they are not developed from the ectoderm or endoderm but arise in another way.

Beginning with flatworms, all higher animals have a mass of cells between the ectoderm and the endoderm, appropriately called the MESODERM ('middle skin'). This layer gives rise to muscles and to other structures which make possible an increasing complexity and efficiency in animal activities. Like almost all characters of animals, the mesoderm does not appear suddenly in fully developed form. Its early beginnings are perhaps represented in the amoeboid mesenchyme cells of the hydra. But we do not consider mesenchyme to be true mesoderm until, as in the flatworms and in all higher animals, it is more massive than either ectoderm or endoderm and gives rise to definite structures, such as the reproductive organs.

Two-layered animals, like the hydra and the obelia, are usually small and fragile, because ectoderm and endoderm are essentially single layers of epithelium. The largest two-layered animals are certain jellyfishes; these have attained great size and a certain amount of body firmness by the secretion of great quantities of a viscous, non-living jellylike material between ectoderm and endoderm. In three-layered animals the mesenchyme has been increased from a scattered group of wandering amoeboid cells to a many-layered tissue that gives firmness and bulk to the body.

The coelenterates were animals organized on the tissue level. From flatworms to man, animals are constructed on a still higher level of organization. Not only do cells work together to form tissues, but tissues of various kinds are closely associated to form one structure, called an ORGAN, adapted for the efficient performance of some one function. Thus the human stomach is an organ composed of epithelium, connective tissue, muscle layers, and nervous tissue. The epithelium lines the cavity and contains the gland cells which secrete the gastric juice; the muscle layers cause the stomach contractions; the nervous tissue co-ordinates the muscle contractions and relates stomach activity to the whole body; and the connective tissue binds the various layers together. An organ usually co-operates with other organs or parts in the performance of some life-activity, and such a group of structures devoted to one activity is termed an ORGAN-SYSTEM. Thus, the stomach is part of the digestive system; and all the other parts of this system, such as the oesophagus, liver, and intestine, are necessary for the proper performance of digestion in man. The higher animals are made up of a number of such organ-systems, as the digestive system, excretory system, circulatory system, nervous system, and so on. Flatworms do not have all of these, and the ones they do possess are not all very well developed; but they are the lowest phylum of animals built on the ORGAN-SYSTEM LEVEL OF CONSTRUCTION.

In the DIGESTIVE SYSTEM of planaria the mouth, curiously enough, is not on the head but near the middle of the ventral surface. It opens into a cavity which contains a tubular muscular organ, the PHARYNX, attached only at its anterior end. The

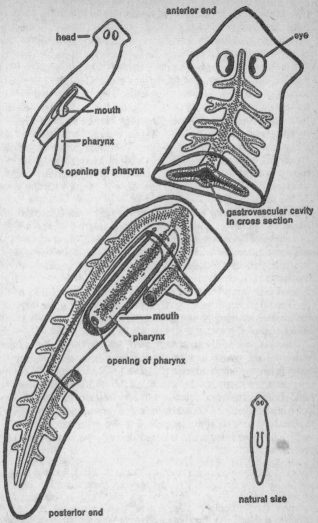

PLANARIA, cut open to show construction of DIGESTIVE SYSTEM, pharynx withdrawn. Small drawing in upper left shows animal with part of body cut away and pharynx extended.

pharynx contains complex muscular layers and many gland cells. By means of the muscles, the pharynx can be greatly lengthened and then protruded from the mouth for some distance; it behaves in this way during feeding. Planarias feed on small live animals or on the dead bodies of larger animals. They can sense the presence of food from a considerable distance by means of sensory cells on the head. They move towards their food, mount upon it, and press it against the bottom by means of their muscular bodies. Struggling prey can be successfully held in this way, especially after it has become entangled in the slimy secretion from the worm. The pharynx is protruded posteriorly through the mouth and onto the food. Sucking movements by the action of the muscles of the pharynx tear the food into microscopic bits and swallow it, along with the juices of the prey.

From the anterior attached end of the pharynx the rest of the digestive system extends throughout the interior of the animal. It consists of one anterior branch which runs forward and two posterior branches which pass backward, one on either side of the pharynx to the posterior end. All three branches of this GASTROVASCULAR CAVITY have numerous and fairly regularly spaced side branches, thus providing for the distribution of the food to all parts of the body. The epithelium of the gastrovascular cavity consists simply of the endoderm and corresponds to the endoderm of the hydra.

There is practically no DIGESTION of food in the gastrovascular cavity of the planaria, such as occurs in coelenterates, for the food is broken up into small particles before it enters the cavity, and is thus ready to be taken up by the epithelial cells in amoeboid fashion and formed into FOOD VACUOLES. The digested food is absorbed and passes by diffusion throughout the tissues of the body. There is only one opening to the gastrovascular cavity; indigestible particles are eliminated through the mouth, as in the hydra.

Experiments on a common species of planaria which was fed on liver showed that, after a meal, all the ingested liver was taken into the epithelial cells in about eight hours, and that three to five days were required for the complete digestion of the food vacuoles so formed. Much of the food was found to be converted into fat, which was stored in the intestinal epithelium, and some remained in certain cells as little spheres of stored protein.

Practically all animals can store food reserves upon which they draw in time of need. A small animal like the amoeba stores very little and, unless it goes into the inactive encysted state, will die after about two weeks without food. Hydras survive much longer periods of starvation. But planarias are peculiarly adapted to go for many months without food while remaining active. During the starvation period they use the food stored in the digestive epithelium, whole cells breaking down. Later they begin to digest other tissues, the reproductive organs usually going first. Externally one can observe only that the worms grow steadily smaller though retaining the same general appearance. A worm starved for six months may shrink from a length of 20 mm. to one of 3 mm. Because of their ability to go for months without food, planarias make ideal household pets for people who are too busy or too absent-minded to keep an 'exacting' animal like a canary.

The region between the outer protective epithelium (ectoderm) and the inner digestive epithelium (endoderm) is filled with various organs surrounded by MESENCHYME in the form of amoeboid cells, many of which are free and move about. The muscle layers already mentioned are embedded in this mesenchyme, and it also contains many gland cells which open to the surface and secrete slime or sticky substances. The gland cells are largely derived from the ectoderm; but the organs, muscles, and mesenchyme are mesodermal.

A new system which was not found in any of the forms already studied is the EXCRETORY SYSTEM. This lies in the mesenchyme and consists of a network of fine tubes which run the length of the animal on each side and open to the surface by several minute pores. Side branches from these tubes terminate in the mesenchyme in tiny enlargements known as FLAME CELLS. Each flame cell has a hollow centre in which beats a tuft of cilia simulating a flickering flame. The hollow centre is continuous with the cavity of the tubules of the system, and the ciliary beat causes a current of fluid to move along the tubules to the pores. Since metabolic wastes seem to be excreted to the outside largely by way of the endodermal epithelium and mouth, the flame-cell system (like the contractile vacuoles of protozoa) appears to function primarily for the regulation of the water content of the tissues.

The planaria has a highly complicated REPRODUCTIVE SYSTEM for sexual reproduction. We saw that in sponges single mesenchyme cells became eggs and sperms, and that in coelenterates

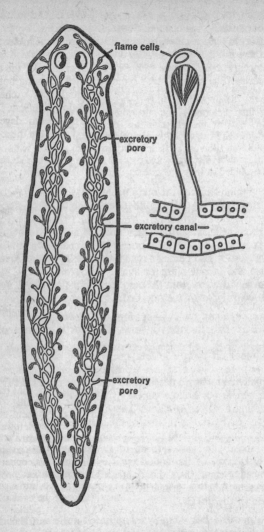

flame cells

excretory pore

excretory canal

excretory pore

EXCRETORY SYSTEM OF A PLANARIA. On the right is shown a single
flame cell attached to a portion of the excretory canal.

mesenchyme cells aggregate into simple ovaries and testes, which discharge their contents, the eggs and sperms, directly to the exterior. In planarias, ovaries and testes arise in the mesenchyme; but there is a system of tubules and chambers in which fertilization occurs, and there are complicated sex organs for the transfer of sperms. The animals are hermaphrodites, forming both male and female sex organs in every individual; but exchange of sperms takes place so that cross-fertilization is effected. After the reproductive season the reproductive system degenerates and is regenerated anew at the beginning of the next sexual period.

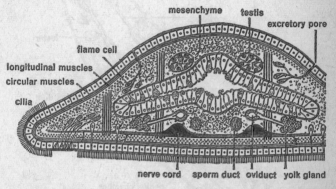

Cross-section through a SEXUALLY MATURE PLANARIA, showing internal organs. Gland cells open to the surface.

When sexually mature, each individual has a pair of OVARIES close behind the eyes. From each ovary a tube, the OVIDUCT, runs backward near the ventral surface. Multiple YOLK GLANDS, consisting of clusters of YOLK CELLS, lie along the oviduct, into which they open. There are numerous TESTES along the sides of the body. From each testis leads a delicate tube, and all these tubes unite on each side to form a prominent SPERM DUCT, which runs backward near the oviduct. The sperm ducts, packed with sperms during the time of sexual maturity, connect with a muscular, protrusible organ called the PENIS, which is used for transfer of the sperms to another planaria. The penis projects into a chamber, the GENITAL CHAMBER, into which there also open the oviducts and a long-stalked sac called the COPULATORY SAC. The genital chamber opens to the exterior by the GENITAL PORE on the ventral surface behind the mouth.

Although each planaria contains a complete male and female sexual apparatus, self-fertilization does not occur; instead, two worms come together and oppose their ventral surfaces. The penis of each is protruded through the genital pore and deposits sperms in the copulatory sac of its partner. After copulation, the worms separate. The sperms soon leave the copulatory sac and travel up the oviducts until they

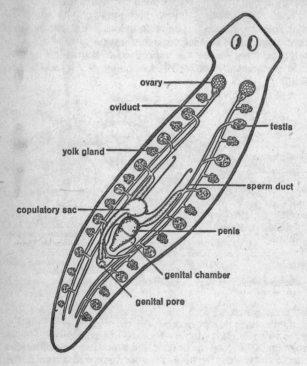

The REPRODUCTIVE SYSTEM OF A PLANARIA includes both male and female sex organs.

reach the ovaries, where they fertilize the ripe eggs as they are discharged. The fertilized eggs pass down the oviducts, and at the same time yolk cells are discharged from the yolk glands into the oviducts. When eggs and yolk cells reach the genital chamber, they become surrounded by a shell to form an egg capsule. The eggs of flatworms are peculiar in that food reserves do not occur in the eggs themselves but are kept in the yolk cells, which accompany the eggs. The capsules (each containing less than ten eggs and thousands of yolk cells) are passed out through the genital pore and are often fastened to objects in the water. They hatch in two or three weeks to minute worms, like their parents except that they lack a reproductive system.

Many planarias have no method of reproduction other than the sexual, but some multiply ASEXUALLY. In this process the worm, without any evident preliminary change, constricts at a region behind the pharynx, and the posterior piece begins to behave as though it were 'revolting' against the domination of the anterior piece. When the whole animal is gliding quietly along, the posterior part may suddenly grip the bottom and hold on, while the anterior headpiece struggles to move forward. After several hours of this 'tug-of-war' the anterior piece finally breaks loose and moves off by itself. Both pieces regenerate the missing parts and become complete worms. Species which have this habit often go for long periods without sexual reproduction, and, in fact, some of them rarely develop sex organs.

ASEXUAL DIVISION. *Left*, just before division. *Right*, just after. The rear piece will soon develop a head, pharynx, and other structures.

In flatworms we see the first appearance of a CENTRAL NERVOUS SYSTEM, the kind of nervous system possessed and further centralized by all higher animals. In the planaria there is in the head a concentration of nervous tissue into a bilobed mass called the BRAIN. From the brain two strandlike concentrations of nerve cells, the NERVE CORDS, run backward through the mesenchyme near the ventral surface. From these ventral nerve cords numerous side branches are given off to the body margins, and the two cords are connected with each other by many cross-strands like the rungs in a ladder. Because of its resemblance to a ladder, this type of system has been called the 'ladder type' of nervous system. The brain and the two cords constitute the central nervous system, a kind of 'main highway' for nervous impulses going from one end of the body to the other. The brain is not necessary for the muscular co-ordination involved

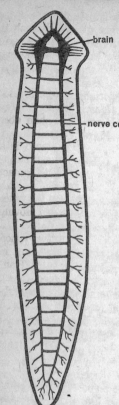

NERVOUS SYSTEM OF A PLANARIA.

in locomotion, for a planaria deprived of its brain will still move along in co-ordinated fashion. It serves chiefly as a sensory relay that receives stimuli from the sense organs and sends them on to the rest of the body. The result is a much more closely knit behaviour than is possible with the diffuse, non-centralized nerve net of the hydra, which lacks definite pathways and a co-ordinating centre:

The nerve net does not disappear in the planaria but persists in addition to the central nervous system. Nerve nets also occur locally in the tissues of almost all higher animals. In man, for example, a well-developed nerve net (connected with the central nervous system) occurs in the wall of the intestine.

Conditions in the external world are conveyed to the nervous system by SENSORY CELLS, slender, elongated cells that lie, with their pointed ends projecting from the body surface, between the epithelial cells. Probably different ones are specialized to receive touch, temperature, and chemical stimuli. Sensory cells are distributed all over the body surface, as in the hydra, but in addition are concentrated in the head to form sense organs. The SENSORY LOBES, pointed projections on each side of the head, are known to be especially sensitive to touch and to water currents, and probably also to food and other chemicals. The two EYES are sense organs specialized for light reception. Each consists of a bowl of black pigment filled with special sensory cells whose ends continue as nerves which enter the brain. The pigment shades the sensory cells from light in all directions but one, and so enables the animal to respond to the direction of the light. Unlike other regions of the ectoderm, that which is immediately above the eye is unpigmented, and thus allows light to pass through to the sensory cells. Planarias whose eyes have been removed still react to light, but more slowly and less exactly than normal worms. This indicates that there must be some light-sensitive cells over the general body surface.

By virtue of abundant sensory cells, specialized sense organs, and a centralized nervous system the planaria shows a more varied BEHAVIOUR and much more rapid responses than does the hydra. Planarias avoid light and are generally found in dark places, under stones or leaves of water plants. If placed in a dish exposed to light, they immediately turn and move toward the darkest part of the dish. They are highly positive to contact and tend to keep the under surface of the body in contact with other objects. They respond to chemical substances in the water and quickly react to the presence of food by turning and moving directly towards it. That is why a piece of raw meat placed in a spring inhabited by planarias attracts hordes of worms, which glide upstream towards the food, guided by the meat juices in the current of water. Planarias react to water currents, and some species regularly move upstream against a current. They also respond to the agitation of the water produced by the animals upon which they prey.

surface ectoderm

pigment

nerve cell

Section through EYE OF A PLANARIA. The eye is sensitive only to light coming towards the open end of the pigment cup. The light-sensitive nerve cells run to the brain. (Modified after Hesse.)

THE flatworms, as illustrated by the planarias, are advanced over two-layered animals in a number of important characters which are possessed by all higher animals. The flatworms are the first animals to have specialized anterior and posterior ends and dorsal and ventral surfaces. They are the first to have a definite head with a concentration of sense

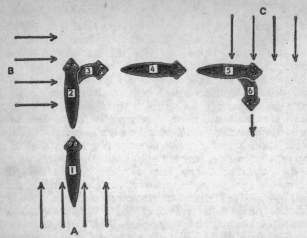

ORIENTATION TO LIGHT. 1–6, successive positions of a planaria. In 1, the animal was moving away from light coming from source A. When it reached position 2, the light was turned off at A and on at B. The worm turned and moved away from the light. At position 5, light B was turned off and C turned on; the worm again oriented away from the light. (Based on Taliaferro.)

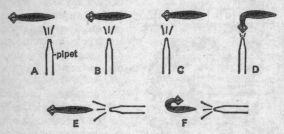

REACTIONS TO WATER CURRENTS produced by a pipet. A, B, there is no response as the current strikes the middle or rear of the body. C, D, the current strikes the sensory lobe on the side of the head and the worm turns toward the current. E, F, the current from the rear passes along the sides of the body to the sensory lobes and the worm turns around toward the current. In nature these reactions orient the worm upstream. (Modified from I. Doflein.)

organs and the development of a central nervous system. And they are the first to make extensive use of a third layer of cells, the mesoderm, which, either by itself or in combination with ectoderm or endoderm, gives rise to organs and organ systems.

CHAPTER 11

The Shapes of Animals

ALTHOUGH animals range in size from microscopic protozoa to massive whales, there are only three basic styles in animal shapes: spherical, radial, and bilateral. A spherical form is assumed by any isolated small quantity of liquid because of the physical forces acting upon it. Small bits of protoplasm or single cells assume this spherical form unless they have a stiff surface layer or skeleton which enables them to maintain some other shape. The amoeba becomes spherical when at rest and must expend energy in extending pseudopods. This type of shape, called SPHERICAL SYMMETRY, is characterized by the arrangement of structures with reference to the centre of a sphere. Since all radii are alike, a spherical animal can be divided into two identical pieces by a cut in any direction through the centre. There is no front or rear, no top or bottom, no right or left sides – at least no permanent ones. This is a disadvantage of the spherical form, since such an animal can show only a very indefinite kind of locomotion – and, in fact, most spherical animals are free-floating. On the other hand, this type of symmetry is admirably suited to the needs of floating animals which do not move under their own power; they cannot swim towards food or away from enemies, but must respond to these on any side from which they may approach. Spherical symmetry is found among adult animals only in protozoa, where it is best seen in certain shelled protozoa (radiolarians, one of which is shown in the drawing at the head of the chapter) which float near the surface of the ocean and feed by means of pseudopods that radiate out in all directions through openings in the shell.

If we imagine a sphere developing a mouth, the surface will no longer be everywhere the same, but will be differentiated into the part bearing the mouth and the part not bearing the mouth. If the food-capturing devices of the animal, such as tentacles, are arranged in a circle around the mouth, this body plan just fits a polyp like the hydra. Both the polyps and the medusas of coelenterates exhibit RADIAL SYMMETRY, in which all radii are alike at any particular level, but there is a differentiation between levels along an axis from the mouth end to the end opposite the mouth. A radial animal may be cut into two identical pieces by a *lengthwise* cut (but not a crosswise cut) through the centre in any direction, for the animal is alike all around the circumference: it has no differentiated sides. This lack of a definite side to go first in locomotion renders moving about somewhat ineffective. Radial animals either drift with the water currents most of the time or live a sessile life. A sessile animal has nothing to 'fear' from below, and the basal end is specialized only for attachment. The circle of tentacles extending from the exposed mouth end is prepared to meet the environment from above and from all sides. Radial symmetry is seen in some protozoa and sponges, but is most characteristic of coelenterates.

For efficient locomotion it is essential that one end should go first. In the planaria there is a definite front or anterior end, which bears the sense organs and which always ventures first into a new environment; and a rear or posterior end, which merely follows along. Such an animal would seem to be open to attack from the rear and from the sides, as compared with a hydra (or medusa), which can detect enemies and ward off their attacks on all sides. However, the concentration of sense organs in the front end enables the animal to detect danger ahead and so better to avoid it. Also, specialization of anterior and posterior ends is related to efficient locomotion, and such animals can better escape from their enemies than can radial animals.

The specialization of the head end is accompanied by a differentiation of the upper and lower surfaces of the body. The undersurface of the planaria bears the mouth and most of the cilia and is quite different from the upper, exposed surface. This type of body form in which there is a difference between front and rear and between upper and lower surfaces is called BILATERAL SYMMETRY. The term 'bilateral' means *two sides* and refers not to these surfaces but to the fact that in these animals the body structures are arranged symmetrically on the two sides with reference to a central plane which runs from the middle of the head end to the middle of the tail end. The paired eyes and sensory lobes of planaria (and the paired structures of man) occur at equal distances to either side of this plane. Single organs are generally located in the mid-line and are bisected by the plane. Bilateral animals have right and left sides, while the hydra, for instance, has no defined sides. A bilateral animal can be cut into two similar pieces by only one particular cut – along the plane which runs down the middle of the body from head to tail and from back to belly. The two resulting pieces are not identical but are mirror-images of each other.

The bilateral symmetry is often imperfect in some degree, owing to the specialization of one side or part over the other. Thus, in man the right arm is usually larger and stronger than the left, and in the brain there is a speech centre on the left side but none on the right. Much more marked asymmetries occur in other bilateral animals, such as the coiling of the snail shell into a spiral (because of unequal growth of the two sides).

When bilateral animals become sedentary, they tend to evolve a modified symmetry which appears superficially like radial symmetry. An example is the starfish, which is said to be secondarily radial or to display secondary radial symmetry.

The bilateral body form lends itself readily to 'streamlining' and, with a head to direct movements, gives rise to fast-moving and therefore very successful animals. Beginning with the flat-worms, all animals (unless secondarily modified) are bilaterally symmetrical.

A form which has no definite symmetry is seen in various protozoa and in most sponges (such animals are said to be *asymmetrical*). However, some kind of symmetry is characteristic of animals in general.

CHAPTER 12

New Parts from Old

REGENERATION, or the ability to repair damage, is almost universal among animals. In man, wounds of considerable size heal and broken bones grow together again, but a lost finger or toe cannot be replaced. Among the invertebrates, the power of repair is much greater, and, in general, the lower the degree of organization of an animal, the greater is its ability to replace lost parts. Such regeneration depends upon the ability of the uninjured cells to produce the kinds of cells destroyed by the injury. Consequently, the more specialized the cells of an animal have become, the less able are they to produce cells different from themselves and so replace the missing parts.

PROTOZOA have a notable capacity for regeneration. Any piece that is not too small will re-form a complete and perfect animal if it contains a nucleus; non-nucleated pieces fail to regenerate. This is not surprising when we recall that in the normal method of reproduction in protozoa the nucleus divides in two and the cell breaks into two parts, each containing half of the nucleus and each capable of growing into a complete individual. In many ciliates there is more than a simple replacement of missing parts. Injury to the basal part of a single large cilium may cause a complete breakdown and resorption of all the cell structures except the nucleus, followed by the differentiation of a new set of parts. Furthermore, even the most complex protozoa lose a certain amount of their specialization at each cell division, and then undergo complete differentiation again before the next division. Thus, their life-histories are fundamentally different from, and their powers of regeneration greater than, those of the many-celled animals whose development involves the progressive specialization of cells through countless cell divisions.

Among the loosely organized SPONGES we saw a very marked ability to regenerate; cells from finely macerated sponges fuse in small masses and develop into complete sponges, as discussed in chapter 6.

COELENTERATES also regenerate very well. Hydroids, like the obelia, have been pressed through fine gauze, after which the separated cells have clumped together to regenerate new polyps. Pieces of the hydra body grow into small but complete hydras.

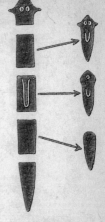

CAPACITY FOR REGENERATION decreases from the anterior to the posterior end. (Based on Child.)

Similarly, some PLANARIAS will regenerate complete worms from almost any piece. The parasitic flatworms, on the other hand, have no ability to replace parts removed; and this statement applies to parasitic animals in general.

Although an earthworm can replace its head, a starfish its arms, and a lobster a leg or an antenna, the HIGHER INVERTEBRATES show, in general, an increased specialization and a corresponding decrease in capacity for regeneration.

Consequently, regeneration has been studied mostly in the protozoa, sponges, hydras, hydroids, and planarias. Planarias, in particular, have been the subject of extensive researches, and certain general facts have been ascertained which apply to all the lower animals.

In the first place, any piece of such animals usually retains the same POLARITY it had while in the whole animal, that is, the regenerated head grows out of the cut end of the piece which faced the anterior end in the whole animal, and the regenerated tail grows out of the cut end which faced the posterior end.

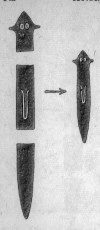

A regenerating piece retains its polarity – a head grows from the anterior end and a tail grows from the posterior end. (Based on Child.)

This *antero-posterior differentiation* is operative throughout the entire animal down to small portions.

Another generalization drawn from these experiments is that the capacity for regeneration is greatest near the anterior end and decreases toward the posterior end. Pieces from the anterior regions of a planaria regenerate faster and form bigger and more normal heads than pieces from posterior regions, and there is a gradual change in these respects along the anteroposterior axis. In some planarias only the pieces from anterior regions are able to form a head, while those farther back effect repair but do not regenerate a head.

The head of a planaria is dominant over the rest of the body, and, in general, any level controls the level posterior to it. One way in which this DOMINANCE can be demonstrated is by means of grafting. If a small bit of the head region of a planaria is grafted into a more posterior level, it will not only grow out into a head but will influence the adjacent tissues to cooperate with it so that a new pharynx, for example, may be formed in the body near this grafted head. If a head piece is grafted into a planaria and then the host's head is cut off, the grafted head may influence the anterior cut surface (which ordinarily would regenerate into a head) to form a tail. In other words, grafts of head pieces reorganize the adjacent tissues into a whole worm in relation to themselves. Grafts from tail regions do not have these effects, but are generally absorbed.

The dominance of the head over the rest of the body is limited by distance. If the animal grows to a sufficient length, its rear part may get beyond the range of dominance of the head. This happens in asexual reproduction of the planaria when the rear part starts to act as if it were 'physiologically isolated' and then finally constricts off as a separate animal. A diminishing of the control of the head over the body is the most important factor in asexual division, as shown by the fact that separation of the rear part can be induced by cutting off the head. And conversely, if the worms are kept in a vaselined dish, so that the posterior subindividual cannot get a good hold on the substratum, the worm is prevented from dividing.

Until the moment when a planaria starts to construct, there is no external evidence of the physiological isolation of a posterior subindividual. Yet its presence can be shown experimentally. Pieces of worms taken from a region just behind the pharynx usually do not develop heads. But shortly behind this is

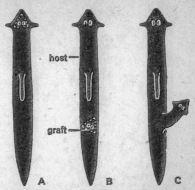

GRAFTING. A, a small piece, indicated by broken lines, is cut out of the head of the donor. B, the graft is placed in a wound made in the body of the host. C, the graft has grown into a small head. (Based on Santos.)

a zone that almost always produces normal heads. This is the region where the worm constricts in asexual division. Apparently this zone represents the developing head of the posterior subindividual. In a longer worm it can be shown that there are three or even four subindividuals, one behind the other, and each indicated by a region which shows an increased ability to produce worms with normal heads.

If the anterior end of a planaria is cut down the middle, and the two halves are prevented from growing together again into a single head by renewing the wound several times, then each half will regenerate the missing parts and a two-headed planaria will result, with dominance divided equally between the two heads. If the cut goes far enough back, each head will influence the formation of its own pharynx.

All these facts indicate that there is some sort of gradation in essential processes along the anteroposterior axis of a planaria, and we refer to this as the ANTEROPOSTERIOR GRADIENT. Similar experiments on hydras and hydroids have yielded similar results, and in these radially symmetrical animals the gradient is highest at the mouth end and decreases gradually toward the base, resulting in a MOUTH-BASE GRADIENT.

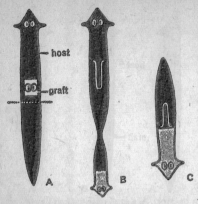

REVERSAL OF ANTEROPOSTERIOR AXIS BY A GRAFT. In this experiment host and graft were of two different species, and the tissues of each remained distinguishable as growth occurred. A, the host pharynx was removed and in its place was grafted a piece of the donor, including the eyes, part of the brain, and adjacent mesenchyme. One week later, the host tail was cut off at the level of the dotted line. B, a few weeks later, the graft has grown out as a small head. The host has developed a new pharynx and is about to divide asexually as indicated by the constriction. C, 74 days after grafting. A pharynx has formed in the host tissue, but it is oriented in a direction opposite that of the old one. A tail has developed at the anterior end of the old host tissue where a head would be expected to develop if the graft were not present. The direction of beat of the host cilia has been reversed. (Based on J. A. Miller.)

Dominance of one region over another is not restricted to animals; *plants* show this phenomenon too, perhaps most diagrammatically in coniferous trees. In a pine tree, for example, the tip of the main stem grows in a vertical direction, and its secondary branches grow out radially. The side branches of the tree are quite different. They grow out horizontally or nearly so, and their secondary branches arise not in a whorl but only from the two sides in bilateral fashion. Now, if the tip of the main stem is destroyed, one of the upper side branches (presumably the one that gets started first) will begin to grow vertically and its secondary branches will grow out in whorls. Since none of the other side branches changes its behaviour in this way, we assume that the other branches are in some way dominated by the new main stem and prevented from exercising their potentiality for vertical growth – unless the new main stem should be injured in some way.

This gradation does not seem to be anatomical in nature, since no definite structural difference is known to exist along the stem of a hydroid, for instance. Consequently it is thought that the gradation is 'physiological', by which we mean that it arises from differences in function rather than from differences in structure. In seeking for the functions that might be the cause of a gradation which affects all regions of the body, we must look to general functions rather than to particular ones. The most general function of the animal body is metabolism – the totality of the chemical processes involved in the building-up and breaking-down of protoplasm. This anteroposterior gradient, according to one theory, consists in a gradation of metabolic rate along the body. The rate is highest in the head region and decreases toward the posterior end.

If the head of a planaria is cut down the middle, each half will regenerate the missing parts. (Based on Child.)

Various lines of evidence have been accumulated in favour of the METABOLIC GRADIENT THEORY. Direct measurements of the oxygen consumption of pieces cut from various levels of the body of planarias have shown that anterior pieces carry on respiration at a faster rate than posterior pieces. Such experiments are not as conclusive as they might seem because of the changes which pieces of a co-ordinated many-celled animal conceivably undergo when cut from the body. If planarias are placed in a poisonous solution of sufficient concentration, they die; but death does not attack all parts at the same time. In these low forms death occurs in a regular progression, beginning at the head end and extending gradually backward. This is explained, on the basis of the theory, by assuming that the parts with the highest metabolic rate are affected first and most severely by a poison, while less active parts are more slowly affected.

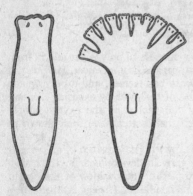

The head of this flat worm was cut repeatedly, and the cut edges not allowed to grow together again. The result is a monster with ten complete heads. (After Lus.)

If the formation of a head at the anterior cut

end of a piece, and of a tail at the posterior cut end, depend on an anteroposterior gradient in that piece, it should be possible to change this result by EXPERIMENTALLY ALTERING THE GRADIENT. Thus, if pieces are cut so short that there is no appreciable difference between the anterior and posterior cut surfaces, such as very short pieces near the eyes in planarias, they regenerate a head at both ends of the piece. Pieces can also be obtained from the posterior regions of planarias which will regenerate a tail at both ends. The gradient can readily be changed in coelenterates. Small pieces of a hydra taken from very near the mouth will grow a mouth and a crown of tentacles at both ends. It is possible to obliterate the gradient in pieces of hydroid stems by putting them in anaesthetics for a time; when replaced in

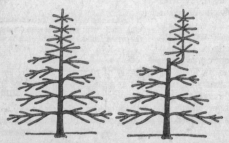

DOMINANCE IN PLANTS. This conifer tree was 'decapitated'. Soon afterwards the uppermost branch grew vertically and became the new main stem. (Based on Child.)

water, such pieces may grow out several polyps irregularly from whatever part of the piece happens to be uppermost. The polarity of pieces of hydroid stems can be reversed, and a polyp made to grow out at the basal end of the piece by exposing this end to a better oxygen supply and decreasing the oxygen supply at the upper end. Other external factors have been shown to operate in the same way.

One may inquire HOW GRADIENTS GET STARTED in animals. It seems that they must arise early in development by the action of external factors on protoplasm. The position of the egg in the ovary is one such external condition. The egg is attached by one end to the ovary and is free at the other end. It is known for a good many eggs that this position determines, or at least is correlated with, the polarity of the egg, that is, which end of the egg is to become the anterior (or the mouth end) of the future animal. The polarity is a property of the cytoplasmic background of the egg and is based upon a gradient, a graduated difference of some kind. This gradient is thought to be related to differences in rate of diffusion of oxygen and other substances in the attached and freely exposed ends. Once established, the polarity continues throughout embryonic development.

It has been found that many experiments on regenerating adult flatworms can be duplicated on developing eggs and embryos. By subjecting eggs and embryos to poisonous solutions of proper concentration and at the right time, the development can be greatly modified and all sorts of curious embryos can be obtained. In

Degeneration in a poisonous solution begins at the head end. (Based on Child.)

general, because such solutions act most severely on the parts highest in the gradient, these embryos show suppression of the head region and the sense organs and often have small heads, reduced eyes or eyes fused into one, and so on.

Such results suggest that some kind of gradient is an important factor in embryonic development and furnishes an underlying pattern which controls the orderly development of normal form and proportion. Thus we might think of the several kinds of symmetry as resulting from differences in the number of gradients which act in development. In the truly spherical animal all radii are alike and there is no one main axis of differentiation, though there is some difference between the interior of the cell and its exposed surface. Such an animal can show only a limited amount of differentiation. In an amoeba all points on the surface are alike in that any point is capable of sending out a pseudopod, but the pseudopod proceeds in a definite direction. By means of chemical indicators it has been shown that in an actively moving amoeba there is a definite, though constantly changing, physiological gradient – highest at the tip of the leading pseudopod and lowest at the opposite end of the animal. Radial and bilateral animals have more permanent axes of differentiation. In radial types the main axis (also called the 'polar axis') is from mouth to base. In bilateral types there are, besides the main anteroposterior axis, two

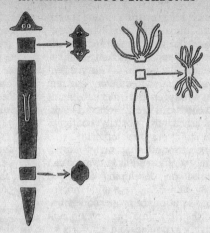

Small pieces regenerate similar structures at both ends.
Left, a planaria. *Right*, a hydra. (Based on Child.)

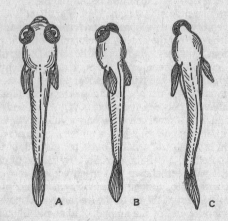

SUPPRESSION OF EYES IN FISH as a result of inhibition of the head by a harmful substance
(in this case magnesium chloride). A, normal individual. B, fish with eyes close together
as a result of inhibition of growth of median region of head. C, fish with single median
eye due to inhibition at the very early stage in head development. (After Stockard.)

minor axes of differentiation. There is evidence for physiological gradients in these two axes, one highest in the mid-region and decreasing on the two sides (the mediolateral gradient), and one higher on the ventral surface and decreasing toward the dorsal surface (the ventrodorsal gradient). These two minor gradients are usually masked by the more prominent antero-posterior gradient. And in higher animals even the anteroposterior gradient is obscured by the complexity of adult structure; it can be clearly shown only in the embryo.

THE experiments on regenerating invertebrates have helped us to understand and have been fruitful in suggesting experimental means of approach to the general problems of animal form, growth, and development.

CHAPTER 13

Free-living and Parasitic Flatworms

LIVING at the expense of one's neighbour is an old habit among animals. Practically all animals harbour one or more kinds of parasites, and most of these are themselves hosts to still smaller parasites. The total bulk of the parasites residing in one host is necessarily less than that of the free-living animal which provides the food and lodging for so many unwelcome guests. But from the standpoint of actual numbers of organisms, the animal kingdom has many more parasitic than free-living individuals.

Nearly every phylum has its parasitic members, and some phyla have more than their share. Of the three classes of flat-worms which compose the phylum Platyhelminthes, two are exclusively parasitic and one consists mostly of free-living animals. The principles of parasitism will be discussed in describing the flukes and tapeworms. But the beginnings of parasitism are to be found in some of the free-living flatworms.

FREE-LIVING FLATWORMS

THE free-living flatworms are much like planarias. Externally they are covered with cilia, the beating of which creates in the water the 'turbulence' that suggested the name of the class, TURBELLARIA. The form of the gastrovascular cavity provides a basis for dividing turbellarians into several groups.

The most primitive group consists of tiny worms which have a mouth but no gastrovascular cavity, and hence are called ACOELS ('without a cavity'). Food is swallowed into a solid mass of endoderm cells and there digested. There is no excretory system. The nervous system, which has several radially arranged main strands of nervous tissue and an anterior sense organ consisting of a cavity containing a hard particle, is similar to that of ctenophores. Acoels are all marine; and because of their

Microstomum, a RHABDOCOEL. Eight subindividuals, each with its own mouth, can be distinguished. (Modified after von Graff.)

small size (usually only about one-tenth of an inch in length), they are difficult to see as they swim or creep about among the rocks and seaweed along shores. They interest us chiefly because they represent a stage of complexity somewhere between that of ctenophores and the more highly developed turbellarians like the planaria.

The RHABDOCOELS ('rodlike cavity') have a straight, unbranched, gastrovascular cavity. They are advanced in structure over the acoels, having a flame-cell system and a more highly developed nervous system. The worms are tiny, usually from one-tenth to a quarter inch in length. They occur in fresh and salt water. *Microstomum* is a rhabdocoel that undergoes asexual reproduction like the planaria. The parts fail to separate at once, so that there result chains of as many as eight or even sixteen subindividuals, each with its own mouth. This animal is interesting also in that it possesses stinging capsules – not of its own manufacture but stolen from the hydras upon which it feeds. The stinging capsules pass from the gastrovascular cavity of the worm through the mesenchyme to the ectoderm. There they become lodged among the epithelial cells, ready to be used in defence.

A LAND PLANARIAN (*Bipalium*) hanging from a branch by a string of mucus as it lowers itself to the ground. Found in greenhouses. (After Kew.)

The group of turbellaria to which the planaria belongs is distinguished by a gastrovascular cavity that has three main branches, one anterior to the attachment of the pharynx and two posterior. They are appropriately called TRICLADS ('three branched'). Besides the fresh-water forms there are also marine and land triclads.

The POLYCLADS, so named from their many-branched gastrovascular cavity, are exclusively marine. They are very thin, leaflike animals – sometimes almost as broad as they are long – and are, in general, the largest of the turbellaria, some species reaching lengths up to six inches. They usually have numerous small eyes and often a pair of sensory tentacles which project from the dorsal surface near the anterior end. Their larvas are free-swimming and have eight ciliated lobes, which are thought by some to indicate a relationship to the eight ciliated rows of ctenophores, though this is very doubtful.

SUCCESSFUL parasitism requires marked adaptation of the parasite to its host, and this evolves slowly. We can distinguish certain intermediate stages in this evolution among free-living flatworms.

Two animals of different species are often found living together in constant association. One derives benefit from the relationship while the other is apparently not injured. Such an association, called COMMENSALISM, is illustrated by *Bdelloura*, a flatworm that lives attached to the gills of the king crab (see chap. 23). Bdelloura is the *commensal* and receives shelter, free transport, and nourishment in the form of tiny scraps from the food of the king crab, its *host*.

Sometimes a commensal incidentally benefits the host, which may then become so dependent upon this service that it cannot get along without the commensal. Such an association of mutual benefit is called MUTUALISM. An example of this has already been given in the case of the intestinal flagellates of termites (p. 57). A well-known case of mutualism is that of *Convoluta roscoffensis*, a tiny marine acoel. The young convolutas are colourless when first hatched, but shortly they take on a green colour as their mesenchyme becomes filled with small, green, plantlike flagellates. After entering the convolutas, the flagellates lose their flagella and undergo other changes, but continue to carry on photosynthesis, producing carbohydrates and oxygen. From the flatworm they receive a sheltered place in the sunlight and a steady supply of carbon dioxide, nitrogen, and phosphorus (from the metabolic wastes of the animal). These last are particularly important because nitrogen and phosphorus, in a form available for protein manufacture, are at a premium in the ocean. While the young con-

(*Hoploplana*), a POLY-CLAD, has a many-branched gastrovascular cavity.

volutas feed like other flatworms, the adults do not feed and are completely dependent for their nourishment upon their plant guests. Experiments have shown that the worms live longer in the light (where the contained flagellates can carry on photosynthesis) than in the dark. Convolutas live on sandy shores between tide-levels and regularly migrate with the tide. As a result, the algae are exposed to sunlight for a maximum time. It should be added that the relationship between Convoluta and its algae has two different phases. In the early stages of the association the worms feed, and there is also a passage of fat from the algae to the tissues of the animal. In the later stages, when the worms stop feeding, they digest the algae in their tissues. This finally results in the death of the worms and suggests that the mutualism is not well balanced.

FREE-SWIMMING LARVA of a polyclad. (After Lang.)

Commensalism usually evolves, not in the direction of mutualism, but towards PARASITISM. A commensal that at first takes only shelter, and then scraps of food, finally begins to feed on the tissues of the host body, and the host suffers a certain amount of harm. Should the parasite become so well adjusted that it causes little damage, and, in fact, finally proves to be of some service to the host, the parasitism becomes

Bdelloura, a TRICLAD that lives as a COMMENSAL on the gills of the king crab.

a mutualism. Thus, the three kinds of relationships are only different stages in the process of living-together, and it is not always possible to draw a sharp line between them. Mutualism and parasitism can probably arise directly from commensalism, but they may evolve from each other. Since mutualism requires the greater number of adjustments, it is relatively rare as compared with parasitism.

A well-adapted parasite usually lives without causing serious harm to its host, for the success of any parasite depends upon the continued success of the host. A well-adapted parasite, already mentioned, is the trypanosome (p. 54), which lives in the blood of African wild game with no apparent ill effect upon the game. The fact that this same flagellate causes a severe illness, and finally death, when it gets into the blood of man or his domestic animals, is taken as an indication that the parasite has only relatively recently come into contact with these hosts. Neither parasite nor host has had time to make the proper adjustments. Some of the flatworms parasitic in man are fairly well adapted – for example, the tapeworms. The flukes are often less so.

FLUKES

THE flukes (class TREMATODA) differ from free-swimming flatworms in the loss of external cilia (in the adult) and in the development of a thick outer layer, the CUTICLE. Flukes have one or more SUCKERS by which they cling to their host. One usually occurs at the anterior end surrounding the mouth, another at the posterior end or on the ventral surface near the middle of the body. The mouth leads through a muscular pharynx into a two-forked gastrovascular cavity. There is an excretory system of flame cells and canals. The nervous system is usually well developed, but there are no special sense organs. In all these respects the flukes are relatively simple animals.

However, when we consider the REPRODUCTIVE SYSTEM, which occupies most of the animal's interior, we find a degree of complexity seldom equalled and not surpassed even in the higher animals. In the posterior part of the body is a pair of testes, from each of which a duct leads to the genital pore near the second sucker. In front of the testes is a single ovary, from which a long, much-coiled tube, the UTERUS, also leads to the genital pore. In the body margins are numerous yolk glands, whose ducts connect with the uterus. The presence of the uterus is the chief difference from the reproductive system of planarias. In the uterus are stored the immense numbers of eggs found in connexion with the parasitic habit.

The *lowest grade of parasitism* is that practised by the flukes which live as EXTERNAL PARASITES attached to the skin or to the gills of fish, feeding upon the epithelial tissue or on blood. Hanging on to the outside of a fast-moving object is no easy matter, and these flukes frequently have an enormously developed sucker (or group of suckers) at the posterior end, besides numerous hooks. As they move about on the surface of their host, these flukes might stray occasionally into the cavities which communicate with the exterior: the mouth, the nasal passages, and the urinary bladder. Thus, it is not surprising to find that many aquatic vertebrates regularly harbour parasitic flukes which have become adapted to live in these cavities, where the danger of being swept off is very much less. The flukes living on the skin and gills and those inhabiting cavities which communicate freely with the exterior have simple life-histories, like those of the free-living flatworms.

Convoluta, an ACOEL that has a dark spinach-green colour from the thousands of green flagellates that live embedded in the mesenchyme. (After Keeble and Gamble.)

The flukes which live as INTERNAL PARASITES, embedded in the tissues or clinging to the lining of cavities far from the surface, have little trouble holding on securely; their hooks or suckers are not as elaborate as those of external parasites. But the problem of getting their offspring established in a new host is a much more difficult one. It has been solved by an increase in the number of potential offspring and by the development of complex life-histories. All the flukes which parasitize man are internal.

The most important of these are the BLOOD FLUKES or schisto-somas, elongated and slender flukes which differ from most in that they are not hermaphroditic but occur as separate males and females. The sides of the male fold over to form a groove in which the longer and more slender female is held. In *Schistosoma japonicum*, the species which we shall consider, the worms live in the blood vessels of the intestine, clinging to the walls of the vessels by means of suckers and feeding on blood.

The female lays her eggs in the small blood vessels of the intestine wall, close to the cavity of the intestine. As the blood vessels of the host become congested with eggs, the walls of the vessels rupture, the intestinal epithelium breaks, and the eggs are discharged into the cavity of the intestine. From there the microscopic eggs are carried out in the faeces. If the faeces were removed by a modern sewage system or deposited in a dry

EXTERNAL PARASITIC FLUKE (*Acanthocotyle*) moving about on the surface of a fish, its host. It holds on by means of the large posterior sucker. (After Monticelli.)

place, that would be the end of the young schistosomas. But in China and Japan, where this parasite flourishes, human faeces are used to fertilize the soil, and for that purpose are conserved in reservoirs on the banks of canals or irrigation ditches. This provides the eggs with ready access to water, where they hatch. A ciliated larva, the MIRACIDIUM, emerges and swims about. If the miracidium does not encounter a snail of a certain kind, it perishes after about twenty-four hours. If it comes into contact with the right kind of snail, it burrows into the soft body of the snail and feeds on the tissues. Meanwhile, the cilia are lost and the miracidium is transformed into a sac, called a SPOROCYST, which produces asexual buds internally. These buds called CERCARIAS, resemble the adults in several ways. They have two suckers, a forked digestive tube, and an excretory system with flame cells, but differ in possessing tails. They make their way out of the snail and swim about near the surface of the water, where they come into contact with the skin of a man who is bathing or wading. Millions of Chinese and Japanese are infected during the planting of rice, as they stand barelegged in the shallow water of the rice fields. The diagram which heads the chapter shows the principal stages in the life-history and the method of infection in man. The cercaria attaches itself to the skin and (by means of glands) digests its way through the skin into a blood vessel. It is carried in the blood stream to the blood vessels of the intestine. There the young fluke feeds and grows into an adult worm, finally mating with another that entered at the same time or with one already established from a previous infection.

EXTERNAL PARASITIC FLUKE (*Gyrodactylus*) holds on to its goldfish host with a large sucker surrounded by hooks. (Modified after Kukenthal.)

PARASITIC FLUKE (*Polystoma*) in mouth cavity of a turtle. (Modified after Stunkard.)

The presence of schistosomas in man causes a disease (schistosomiasis) characterized by body pains, a rash, and a cough in the early stages, severe dysentery and anaemia later on. Victims may live for many years but gradually become weak and emaciated and eventually many die of exhaustion or succumb to other diseases because of their weakened condition.

Control measures for schistosomiasis might reasonably begin with sanitary disposal of human faeces. This is not practicable in the Orient, however, for the use of human faeces as fertilizers is an important part of the economy of the people. It has made possible intensive cultivation of the same soil for thousands of years (whereas in the United States

HUMAN BLOOD FLUKE (*Schisto-soma japonicum*). (Modified after Looss.)

the nitrogen content of the soil is some-times depleted in two or three genera-tions). In Japan, where infection is res-tricted to small areas, it has been possible to kill the snails with chemicals. In China, where infection involves vast areas in the Yangtze Valley, it is better to educate the farmers to conserve the faeces for a few weeks before using it in the fields, so that the young schisto-somes within their protective egg mem-branes will have died.

Other species of blood flukes infest peoples in the north-eastern part of South America, certain places in western Asia, and a large part of Africa. In Egypt the commonest blood fluke lives in the blood vessels of the urinary tract; the eggs are extruded into the bladder and then come out in the urine. The life-history involves a snail. Infection occurs during bathing or from drinking water containing cercarias. The disease is spreading, owing to increasing use of irrigation in farming. It has been estimated that about three-fourths of the population is infected, and one parasitologist has said that Egypt will never be a country of consequence until there is better control of the flukes that are now draining the energy of the people.

WHEN a parasite lives for part of its life-cycle in one kind of animal and spends another part of its life-cycle in another kind of animal, the host that harbours the sexually mature form is said to be the FINAL HOST, and the one that harbours the young stages is called the INTERMEDIATE HOST. In the case of *Schistosoma japonicum*, man is the final host, and a certain species of snail is the intermediate host.

A similar type of life-history is shown by the large LIVER FLUKE, *Fasciola hepatica*, which inhabits the bile passages of the liver in cattle and sheep and inflicts severe, often fatal damage – with important economic consequences for stock-raisers. Instead of boring directly into the final host, after leaving the snail, the cercarias of this fluke encyst on grasses and other vegetation and are eaten by the final host.

A liver fluke that lives in man and illustrates a life-cycle involving *two intermediate hosts* is the CHINESE LIVER FLUKE,

Clonorchis (Opisthorcis) sinensis, of China, Japan, and Korea. The adult is about half an inch long and has two suckers, one at the anterior end and one a short distance behind this. The thick cuticle is resistant to digestive fluids. The fluke is hermaphroditic, and the fertilized eggs pass from the liver into the intestine and out with the faeces. If the faeces get into water, as they commonly do, the eggs do not hatch into free-swimming miracidia, as in most flukes, but are eaten by snails. Within the digestive

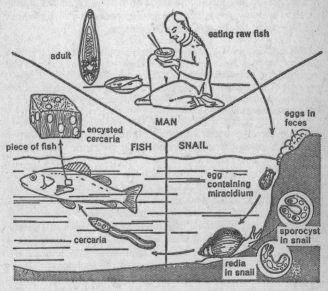

adult

eating raw fish

encysted cercaria

piece of fish

MAN

FISH   SNAIL

eggs in feces

egg containing miracidium

cercaria

sporocyst in snail

redia in snail

LIFE-CYCLE OF THE CHINESE LIVER FLUKE (*Clonorchis*). (Based on Faust.)

tract of the snail the egg opens, and the miracidium emerges and makes its way through the wall of the digestive tract into the tissues of the snail. There it becomes transformed into a sporocyst, which produces, instead of cercarias, another asexual form, the REDIA. The development of redias further increases the number of young forms, for each redia subsequently produces many cercarias, which escape from the snail and swim about. The cercarias encyst, not on grass like those of the sheep fluke, but in the muscles of a fish, which thus serves as the second intermediate host. They burrow through the skin of the fish, lose their tails, and secrete about themselves protective capsules. The fish responds by forming an outer capsule around each one produced by a parasite. There they remain until the fish is eaten by the final host, man. In the human stomach the cysts are digested out of the flesh, and in the intestine the capsule is weakened and the young fluke emerges. It makes its way up the bile duct and into the smaller bile passages of the liver, where it attaches by its suckers and feeds on blood. These flukes may persist for many years, causing serious anaemia and disease of the liver from blocking of the bile passages.

In this case control should be a relatively simple matter, for it is only necessary to cook fresh-water fish thoroughly to destroy the encysted cercarias. Yet in certain regions in the south of China from 75 to 100 per cent of the natives are infected. Not only do they like to eat raw fish, but, unfortunately, the cost of the fuel necessary to cook the fish is an economic problem.

As they become better adapted, parasites usually lose active means of transport and depend upon more or less PASSIVE TRANSFER to new hosts. In the flukes and other parasites passive transfer is frequently achieved through the food habits of the host. Thus the sheep liver fluke is rare in man because man does not usually eat grass, though he sometimes becomes infected by eating watercress upon which cercarias have encysted. He frequently gets the Chinese liver fluke, however, because in some places he habitually eats fresh-water fish containing encysted cercarias. Having once smuggled themselves into a host, the 'problem' of an internal parasite (not of the individual animal, but of the species) is how to get the offspring into new hosts. The easiest way of leaving a host is with the outgoing faeces, just as the easiest way of entering is by way of the mouth. But passive modes of transfer are very hazardous. The chance of having the eggs or the larvas eaten by the right kind of host at the proper stage in the life cycle is very small. Only parasites which produce enormous numbers of potential young can survive this kind of life-cycle. It is not surprising, therefore, that most parasites seem to live only to reproduce, the reproductive organs occupying most of the animal's body.

The chance that an egg will reach a suitable spot for hatching is remote, and the probability that the miracidium will find a snail within a short time (the time being limited by the small amount of energy available in the food stored within the egg) is even more remote. But if even a single miracidium manages to enter a snail, it can multiply within its intermediate host by asexual means and so compensate for the enormous loss of potential individuals by the random distribution of the eggs. It has been estimated that a single miracidium will give rise, through several generations of sporocysts and redias, to as many

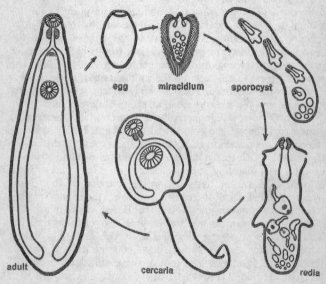

Typical stages in the LIFE-CYCLE OF LIVER FLUKES. The adult is usually ½–1 inch long: the other stages are microscopic. (Modified after Faust.)

as ten thousand cercarias. Of all these, only a few will reach new final hosts.

PARASITISM often results in weakness or disease of the host, but the EFFECTS OF PARASITISM ON THE PARASITE are even more marked. The parasite retains the general plan of organization of its phylum, but it becomes so completely adapted to its peculiar environment that it usually loses many of the structures characteristic of its free-living relatives. In adult flukes we saw a loss of external cilia and of sense organs, structures related to locomotion. In even more highly adapted parasites, like the tapeworms, there is a loss of still more structures.

TAPEWORMS

THE tapeworms (class CESTODA) are usually long, flat, ribbon-like animals, some species of which live as adults in the intestine of probably every species of vertebrate.

The most common tapeworm of man is the so-called 'beef tapeworm' (*Taenia saginata*). It maintains its place in the intestine, despite the constant flow of materials, by means of four suckers on the minute knoblike HEAD (or scolex). Behind the head is a short neck or growing region, from which a series of body SECTIONS (proglottids) are constantly budded off. The sections closest to the neck are the youngest ones; those farthest away, the most mature. Thus the body widens gradually along its length, and the sections are in all stages of development.

The body is covered externally by a protective cuticle, as in flukes; and there is no ectodermal epithelium. (After secreting the cuticle the ectoderm cells sink into the mesenchyme.) Unlike the flukes, which feed actively on the tissues of their host and do their own digesting, tapeworms have no mouth and *no trace of a digestive system*. They live in the intestine of their host, where digested food is readily available; there they simply 'soak up' their nourishment – truly the laziest way of living.

Beneath the cuticle are longitudinal and circular muscles. The NERVOUS SYSTEM is like that of turbellarians and flukes, but less well developed. From a small concentration of nervous tissue in the head two longitudinal nerve cords run backward through the body. Between the nerve cords, and parallel with them, run two longitudinal EXCRETORY CANALS, connected with each other by a crosswise canal near the posterior border of each body section. The smaller branches of the excretory system end in flame cells.

The REPRODUCTIVE SYSTEM lies embedded in the mesenchyme and is so highly developed in mature sections that the tapeworm is sometimes described as nothing but a bag of reproductive organs, a complete set of which, both male and female, develops at some time in every section. The male sex organs start to grow first. They consist of numerous small TESTES, scattered throughout the mesenchyme and connected by many fine tubes

with a single large convoluted SPERM DUCT, the end of which
is modified as a muscular organ, the penis, for the transfer of
sperms. The sperm duct opens into the GENITAL CHAMBER, which
connects with the outside through a GENITAL PORE. Running
parallel with the sperm duct and also opening into the genital
chamber is the VAGINA, a female duct which receives sperms.
Self-fertilization can occur within the same segment, or cross-
fertilization can take place between the segments of different
worms when two or more are present in the same host. But the

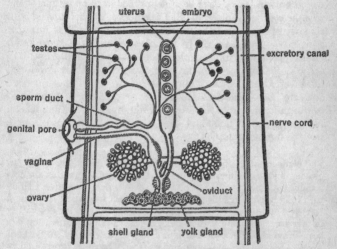

TYPICAL SECTION OF A BEEF TAPEWORM (*Taenia saginata*.)

most common method is transfer of sperms from one section of
the worm to a more mature section farther down the length of
the same worm. This is possible when the animal is folded back
on itself for part of its length. As the section matures, the male
sex organs degenerate while the female sex organs develop.
Eggs are produced in a pair of OVARIES and pass into the OVIDUCT.
There they are fertilized by sperms that have entered through the
vagina and have been stored in an enlarged portion of its inner
end.

The fertilized eggs are combined with yolk cells from the
YOLK GLAND and are then covered with a shell secreted by the

SHELL GLAND. The completed eggs pass forward into the UTERUS, which at first is a single sac (as shown in the diagram) but later develops numerous branches. Eventually all the female organs degenerate except the uterus, which becomes enormously distended with eggs that are already undergoing development into embryos. At this stage the 'ripe' sections, each containing many thousands of young embryos, detach from the worm and pass out with the faeces.

Transfer to a new final host is entirely passive and involves the eating habits of man and cow. The cow eats vegetation on which human faeces have been deposited. In the intestine of the cow the eggshell is digested off. The embryo, which is armed with six sharp hooks, bores its way through the wall of the intestine and into a blood vessel. In the blood stream the SIX-HOOKED EMBRYO is carried to a muscle. There it remains and grows into a sac or BLADDER, from the inner wall of which is developed the inverted head of the future tapeworm. When man eats raw or 'rare' beef, the enclosed bladder is digested off; the head everts and attaches to the intestinal wall by means of its suckers. Nourished by an abundant food supply, it soon grows a long body and produces eggs.

The bladders of the beef tapeworm occur most frequently in the jaw muscles and in the muscles of the heart; these are the parts of the cow usually examined by meat inspectors. The bladders are almost half an inch long but can readily be overlooked. Meat inspection in Western countries and in the United States has greatly reduced the occurrence of this once common parasite. But there are still many cases, and it is best to avoid eating beef that is not cooked thoroughly. In parts of Africa where sanitation is poor, and in Tibet, where beef is prepared by broiling large pieces over an open fire, a large proportion of the population is infected. Among the Hindus of India, who consider the cow sacred and have religious restrictions against eating beef, to be caught with a beef tapeworm would undoubtedly prove embarrassing.

Likewise, the pork tapeworm (*Taenia solium*) is rare among Jews and Mohammedans, who avoid the meat of the hog, and very common in parts of Europe, where pork is eaten without thorough cooking. Meat inspection has made it uncommon in the United States. The pork tapeworm resembles the beef tapeworm closely and has a similar life-history, except that the bladders develop in pigs. It is especially dangerous, however, because self-infection with the embryos can occur, and the bladderworms then develop in man. If these settle in the muscles no great harm results. Sometimes, however, they lodge and grow in the eyeball, interfering with vision. Certain cases of insanity or epilepsy are really due to bladderworms in the brain.

A very thin person is frequently accused by his friends of harbouring a tapeworm, and it is true that infected individuals are sometimes emaciated. The anaemia and the nervous disorders that sometimes occur are not due to loss of food but to the poisonous substances given off by the parasite. Also, the mere bulk of the worm, especially when folded back on itself many times, may block the intestine and produce serious temporary illness.

The presence of a tapeworm can be detected by the appearance in the faeces of the white, ripe sections loaded with embryos. The only way to get rid of the parasite is to take by mouth some drug which kills the head and causes it to detach from the intestinal wall, whereupon the whole worm is evacuated with the faeces.

Many tapeworms have more than one intermediate host. The 'broad fish tapeworm' (*Dibothriocephalus latus*), which may be three-quarters of an inch wide and 60 feet long, is the largest and the most injurious tapeworm that lives in man. The life-history requires that the eggs reach water, that the larvas which hatch are eaten by copepods (small crustaceans), and that the copepods are eaten by fish. Man gets the parasite when he eats raw or imperfectly cooked fish.

The fish tapeworm occurs in many places all over the world and has been known for centuries in the Baltic region of Europe, where in some localities nearly all the people are infected. In relatively recent years Baltic immigrants to the Great Lakes region have brought these tapeworms with them and have established them by infecting the fish in the lakes of Minnesota, northern Michigan, and Canada. Since these lakes supply millions of pounds of fresh fish to other parts of the country, and since visitors to the region carry tapeworms home to their own localities, this parasite is spreading in the United States. Pikes and pickerels are the fish most commonly infected with the bladderworms, but others may be as well. Those who eat fish from these regions should never taste the raw fish during the preparation and should be careful to cook the fish very thoroughly; smoked fish may not be safe.

SOMETIMES MAN IS THE INTERMEDIATE HOST for a tapeworm that lives its adult life in some other mammal. *Echinococcus granulosus* is a minute tapeworm (with only three or four sections) that lives as an adult in the intestine of the dog and occurs only as a larva in man. In spite of the small size of the adult, the larva is enormous. Human infection results from drinking contaminated water or from allowing dogs to lick the face and hands.

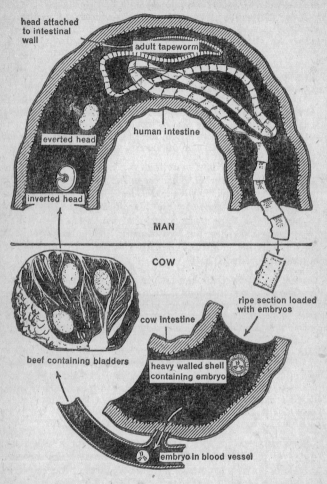

head attached to intestinal wall

adult tapeworm

everted head

human intestine

inverted head

MAN

COW

ripe section loaded with embryos

cow intestine

beef containing bladders

heavy walled shell containing embryo

embryo in blood vessel

LIFE-CYCLE OF BEEF TAPEWORM (*Taenia saginata*). All stages are shown about natural size, except the six-hooked embryo which is microscopic.

Because of the unclean habits of dogs their tongues are likely to carry tapeworm eggs. The young larva develops into a hollow bladder. From the inner walls of this grow smaller bladders, and within each of these are produced numerous heads. The whole structure is known as a 'hydatid cyst' and may grow to the size of an orange or even larger. When such a cyst develops in the brain, the results are extremely serious. Some cases of epilepsy are due to hydatid cysts. From the 'point of view' of the parasite, development of the cysts in man is unfortunate because man is rarely eaten by dogs, and the cysts cannot reach their final host. The most common intermediate hosts are sheep and cattle. The parasite is common in the great cattle- and sheep-raising regions of the world, of which the United States is one.

PORK TAPEWORM (*Taenia solium*). A, six-hooked embryo. B, bladderworm. C, head of adult showing suckers and hooks. (After various sources.)

No one has ever seen a free-living animal evolve into a parasite. But there are so many animals that lead lives which are transitional between these two extremes that we feel fairly safe in hazarding the guess that parasitism starts as a harmless association or commensalism in which one animal takes shelter in the home of, or on the body of, another. Next, the commensal takes small scraps of the food of its host. When it begins to feed on the tissues of the host, the commensal becomes a parasite. External parasites are little changed from their free-living relatives except for the development of hooks or suckers for holding on. Internal parasites usually show marked structural adaptations to their special environment. The nervous and muscular systems, so important for an active free-living life, may become degenerate. Many highly adapted parasites lose the free-swimming young stages and depend entirely upon passive transfer from host to host. In some intestinal parasites digestive organs are reduced and in others are lost altogether. On the other hand, the reproductive system of parasites is so highly developed that most of the energy of these animals is directed toward one main activity – the production of tremendous numbers of eggs to offset the losses incurred in the hazardous transfer from one host to another. Asexual reproduction within the body of an intermediate host is another method for increasing the number of young forms, and therefore the chance that a reasonable number will reach the final host.

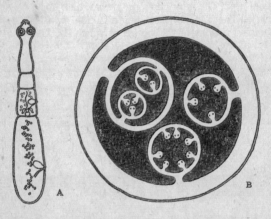

MAN IS THE INTERMEDIATE HOST for *Echinococcus*. A, adult, ⅛ to ¼ inch long, lives in the dog. B, cyst from liver of man. (Modified after Leuckart.)

The young stages of many parasites live embedded in such tissues as liver, muscle, brain, etc. But the adult, which produces the eggs, must live in or near some cavity which has direct access to the outside. The digestive cavity is the one most frequently occupied, since the digestive tract is the easiest and consequently the most popular highway for the entrance and exit of parasites, particularly for those that depend upon passive transfer.

The relation of a parasite to its host requires not only marked adaptation on the part of the parasite but often also an adjustment on the part of the host. The host may secrete a capsule about the larva embedded in its muscles; this helps to confine the activities of the larva. Or the host may develop an immunity to the toxic substances given off by the adult.

The *parasite-host relationship is usually specific.* Some parasites can live in a variety of closely-related hosts, but most can develop in only one particular species. Some can grow in species other than their normal hosts; but when they do so, there is a lack of mutual adjustment and the host or parasite may suffer unduly.

There is a tendency among most people to look upon parasitism as an aberrant way of life and upon parasites as being somehow 'immoral' or at least less 'respectable' than their free-living relatives. But since there are more parasites than free-living individuals, a parasitic existence must be considered a 'normal' way of life. Who can say that the parasite, the very existence of which depends upon doing as little harm as possible to its host, is a less 'considerate' creature than the voracious carnivore, which kills its victim outright?

CHAPTER 14

One-way Traffic – Proboscis Worms

THE proboscis worms are common along seashores under stones and among seaweeds; a few live in fresh water or damp soil. Their elongated flattened bodies range in length from less than an inch to many feet, and they are often coloured a vivid red, orange, or green, with contrasting patterns of stripes and bars. Their most distinctive character is the PROBOSCIS, a long, muscular tube which can be thrown out to grasp prey. The aim of the proboscis is said to be 'unerring', and this is the meaning of NEMERTEA, the name of the phylum to which these worms belong.

Like planarias, they are bilaterally symmetrical. The anterior end is more or less marked off as a head and bears numerous simple eyes and specialized sensory cells. The proboscis worms are not a very large group; they are not usually seen by visitors to the seacoast, nor do they have special economic or medical importance. They are described here because they are the lowest animals to possess two important advances in construction over the flatworms. These structural improvements are present in all higher animals and are, apparently, essential to any increase in complexity over the flatworm plan.

First is an increased efficiency of the DIGESTIVE SYSTEM. The mouth is near the anterior end and opens into a long, straight INTESTINE having short side branches. This intestine extends the length of the body and at the posterior end communicates with the exterior through an opening called the ANUS. This is the first phylum of animals encountered in which all of the members have a digestive tract with two separate openings, one at the anterior end exclusively for taking in food, and the other at the posterior end for the exit of indigestible materials. This 'one-way traffic' has great advantages over the general 'traffic jam' in which the food finds itself in the gastrovascular cavity in coelenterates and flatworms, where the newly ingested food becomes mixed with partly digested food and indigestible residues. In a gastrovascular cavity the lining cells must be both digestive and absorptive. The whole system is inefficient and an obstacle to differentiation of the digestive epithelium. In a one-way system the food passes along a digestive tract which can be differentiated into various parts or organs with specialized functions: one region for food intake, another for digestion, still another for absorption, and so on until the end,

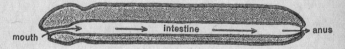

mouth   intestine   anus

Diagram of the DIGESTIVE SYSTEM of a nemertean.

the anus, which is for elimination. All animals above flatworms have an anus, which makes possible a continuous digestive canal through which food passes in only one general direction.

The proboscis of nemerteans, which is often as long as the body, lies inside a muscular sheath just above the digestive tract. When a likely victim approaches, such as a nereis (an annelid worm), the proboscis is extended quickly, wraps around the prey, and entangles it with the aid of a sticky mucous secretion. In some nemerteans the proboscis bears a sharp stylet which pierces the body of the prey and makes a wound into which is poured a poison from glands in the proboscis. The proboscis is fastened to its sheath near the anterior end and is turned inside out when protruded. It is drawn back by a retractor muscle attached to its posterior end.

The second important structural advance pioneered by the nemerteans is a new system, the CIRCULATORY SYSTEM, which takes over the circulatory function of the old gastrovascular cavity. This makes it possible for the intestine to become more efficient in digestion, and leaves the distribution of food and oxygen and other substances to a system more fit for the job. The circulatory system of nemerteans consists of three lengthwise muscular tubes, the BLOOD VESSELS. These lie in the mesenchyme, one on each side of, and one just above the intestine, and connect with each other by transverse vessels. They contain a fluid, the

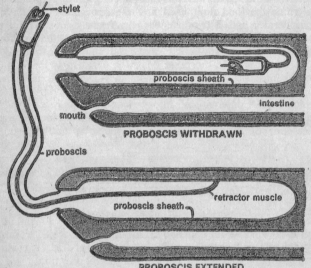

Diagram of the anterior end of a nemertean showing PROBOSCIS withdrawn and extended. When the stylet is lost, it is replaced from a sac of extra stylets in the tip of the proboscis. (Based on Coe.)

Diagram of the CIRCULATORY SYSTEM of a nemertean.

BLOOD, which is generally colourless and contains cells. In some species the cells are red from the same substance (haemoglobin) that colours human blood red. Haemoglobin combines readily with oxygen, making the blood more efficient as an oxygen-carrier. It is most characteristic of vertebrates but occurs in many groups of invertebrates, of which the lowest are the nemerteans. The circulation of nemerteans is primitive in several important respects. There is no special pumping organ or heart to circulate the blood and so move materials from one end of the animal to the other; this is done only by the general movements of the body. Also, the blood vessels are not finely branched; therefore materials must move longer distances to and from cells by the slow process of diffusion. Like most structures when they first arise, the circulatory system of nemerteans lacks specialization and does not do as good a job as the more complex circulatory systems of higher groups.

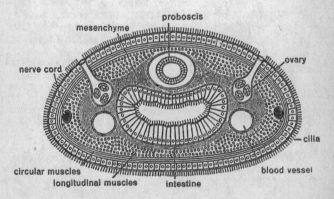

GENERAL BODY STRUCTURE of a nemertean shown in cross-section. (Based on W. R. Coe.)

The GENERAL BODY CONSTRUCTION of nemerteans, apart from the two new features described above, is very similar to that of a planaria and includes all the same organ-systems. The animal is covered completely with a ciliated epithelium which contains many gland cells. Under the epithelium are thick muscular layers, circular and longitudinal, by which the highly contractile animal can execute agile movements of all sorts. The digestive tract consists of an epithelium. Some muscle cells are usually associated with it, but food is moved along chiefly by means of muscular contractions in the body wall which pass down the animal from front to rear and force the food along as they press on the intestine. These muscular waves also assist the flow of blood. Between the gut and the body wall there is a thick layer of mesenchyme cells. Embedded in the mesenchyme is the circulatory system, already described, and also the excretory system, consisting of a pair of lateral canals with side branches ending in flame cells. Wastes are removed from the mesenchyme and from the blood and pass into the canals, which open on the surface by pores. The nervous system is similar to that of flatworms, but the brain is more massive and forms a ring round the digestive tract; longitudinal nerve cords run the length of the worm on each side.

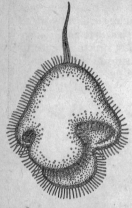

PILIDIUM LARVA.
(Based on C. Wilson.)

The eggs and sperms are produced in little sacs which lie in the mesenchyme between the branches of the intestine. Each opens directly to the outside through its own pore on the surface. There is none of the complicated sexual apparatus seen in flatworms, the sex cells being simply shed to the outside, and in this respect the nemerteans are less specialized than the flatworms. In marine nemerteans there is a ciliated larva shaped like a helmet with earlaps and known as the PILIDIUM LARVA. It has a ventral mouth but no anus, and at the end opposite the mouth is a sense organ topped by a tuft of long stiff flagella. These features suggest that the nemerteans probably arose from an ancestral stock related to the flatworms and ctenophores. The pilidium is similar to the trochophore larva of several of the higher phyla. By a series of rather complicated changes the pilidium larva develops into the adult worm.

CHAPTER 15

Roundworms

MOST people are host at some time or other to the cylindrical white worms called 'roundworms'. Of the fifty different species of roundworms that have been found in man, only about a dozen are common parasites. Of these, some are harmless and do not even make their presence known, while others cause mild or very serious diseases.

Many roundworms look like animated bits of fine sewing-thread; and from the Greek word for thread, 'nema', has been derived the technical name of the group, phylum NEMATODA.

This is by no means a small or obscure phylum. Nematodes are so abundant that a spadeful of garden soil teems with millions of them. When we see a sick dog, our first guess is that he has more roundworms than he can stand. And even in the best-regulated cities roundworms occur in the drinking-water. The widespread occurrence of this group inspired a leading student of the nematodes to write:

If all the matter in the universe except the nematodes were swept away, our world would still be dimly recognizable, and if, as disembodied spirits, we could then investigate it, we should find its mountains, hills, vales, rivers, lakes, and oceans represented by a film of nematodes. The location of towns would be decipherable, since for every massing of human beings there would be a corresponding massing of certain nematodes. Trees would still stand in ghostly rows representing our

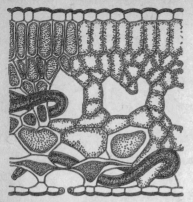

NEMATODES PARASITIC IN PLANTS cause untold damage. As small larvas, they enter through the breathing pores of the leaves. They suck up the cell sap, causing a wilting and withering of the leaves, stunting the plant, and sometimes inducing the growth of galls. The worms shown here (*Aphelenchus*) are in a section of the leaf of a dahlia. They have already injured the cells to the left. (After H. Weber.)

streets and highways. The location of the various plants and animals would still be decipherable, and, had we sufficient knowledge, in many cases even their species could be determined by an examination of their erstwhile nematode parasites.

Some roundworms occupy very specific niches. One species occurs practically only in the appendix of man. Another has never been found anywhere except on the felt mats on which Germans set their mugs of beer. However, the group as a whole lives anywhere that other animals can live. Any collection of earth or of aquatic debris from an ocean, a lake, a pond, or a stream, if examined with a lens, will reveal the tiny white worms thrashing about in a way so characteristic of roundworms that it immediately identifies them.

Nematodes are so remarkably alike that a description of an ASCARIS roughly fits almost any other roundworm. The elongated cylindrical body is pointed at both ends. Curiously enough, it is entirely devoid of cilia, outside and in. The body is covered with a thick, tough CUTICLE secreted by the underlying ectoderm, which has many nuclei but is lacking in cell walls and is called, therefore, a SYNCYTIUM. Under this is a LONGITUDINAL MUSCLE LAYER, divided up into four lengthwise bands by four projections of the syncytium. These bands can be seen on the outside as light lines – a dorsal, a ventral, and the slightly more prominent right and left LATERAL LINES. The muscles are primitive and consist of large length-wise cells with bulblike cytoplasmic expansions which project into the interior. The stiff cuticle and the lack of circular muscle fibres permit bendings of the body in only the dorsoventral plane; and as a result, nematodes

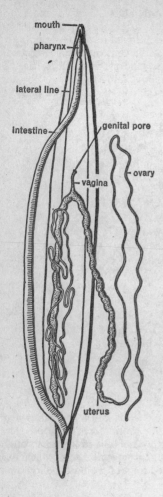

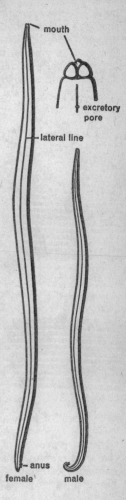

DISSECTION OF A FEMALE ASCARIS. The animal has been slit open along the mid-dorsal line, revealing the internal organs.

ASCARIS, taken from the intestine of a pig. The upper figure on the right shows the anterior end, enlarged.

move in a very erratic and apparently inefficient manner by simply thrashing about. When the worms are free in the water, the whiplike contortions of the body do not result in locomotion; but when they are in the soil, in the contents of the intestine, or in the tissues of the body, the solid particles afford friction and the worms manage to move along fairly well.

The MOUTH is at the anterior tip, encircled by SENSE ORGANS

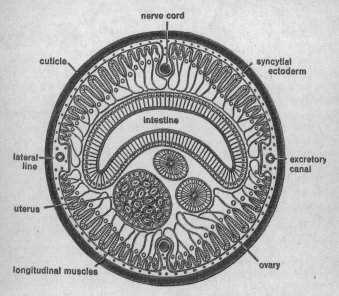

CROSS-SECTION OF A FEMALE ASCARIS.

in the form of protuberances, and leads into a short muscular PHARYNX by means of which the worm sucks in food. The INTESTINE is made up of only one layer of cells and opens through the ANUS near the posterior tip. In parasitic forms which live constantly bathed by the digested food of the host the intestine has no digestive gland cells.

Between the intestine and the muscular layer there is a fluid-filled space formed by the coalescing of vacuoles in a syncytial mesenchyme. The fluid in this space, propelled by the movements of the body, aids in distributing food and oxygen. In higher animals this space develops in a different way, with the result that it is completely lined by mesodermal epithelium and becomes a new cavity called the 'coelom'.

The SPERM OF ASCARIS is amoeboid rather than flagellated, as in most animals.

The NERVOUS SYSTEM consists of a ring of nervous tissue, around the pharynx, from which longitudinal trunks run backward. The principal trunks are dorsal and ventral, and run in the dorsal and ventral thickenings of the ectoderm. The LATERAL LINE, a thickening on each side, contains a canal which is probably excretory, although there are no flame cells, as in flatworms. The two canals unite near the anterior end to form a single tube, which opens ventrally by an EXCRETORY PORE.

The REPRODUCTIVE SYSTEM lies in the space between the intestine and the muscle layer. The sexes are separate, and the males are usually smaller than the females. The reproductive system of each sex consists of a long tube – single in the male, paired in the female – which coils back and forth in the body space. The two OVARIES are long slender tubes which gradually widen into OVIDUCTS and finally into large tubes, the UTERUSES, where the eggs accumulate. The two uteruses unite into one short duct, the VAGINA, which leads to the female GENITAL PORE, situated in the anterior part of the worm on the ventral side. The TESTIS consists of a long coiled tube in which the sperms are formed. It gradually enlarges into a SPERM DUCT, which opens near the posterior end. The sperms are transferred to the female with the aid of a pair of horny bristles.

OF the parasitic nematodes, one of the largest is *Ascaris lumbricoides*, which inhabits the human intestine. The adult is generally several inches to more than a foot in length. The males are smaller than the females and can be distinguished readily by their curved posterior ends. The oval eggs are easily recognized, in microscopic examination, by their warty shells. Two hundred thousand may be laid daily by each female. They pass to the exterior in the faeces, are deposited on the ground, and there develop inside their shells into little worms which are infective to man when the eggs are swallowed. With so many eggs it would seem that everybody would be infected; but hazards

The EGG OF ASCARIS has a shell so resistant that the embryo may continue to develop for a time when placed in concentrated solutions of poisonous chemicals.

occur which are fatal to the developing worm, such as drying, temperatures that are too low (below 60° F.) or too high (above body temperature), or sanitary laws and habits of men. On hatching, the little worms do not remain in the intestine but take a sort of 'tour' through the body. They burrow through the intestine into the blood vessels and are then carried about to various organs, but take hold only in the lungs. They bore through the lung tissue into the bronchial (air) tubes, ascend into the mouth, and are then swallowed back into the stomach, whence they pass again into the intestine. There they remain, growing rapidly to adult size and feeding upon the digested food of their host. They resist digestion themselves by secreting a substance which counteracts the action of the host's enzymes, but if the worms die they are digested by their host.

The greatest damage to the host is done during the migrations of the young; the adult worms in the intestine seem to be relatively harmless unless they occur in large numbers. Up to five thousand worms have been found in one host, but even a hundred worms may block the intestine completely and cause the death of the host. Sometimes they wander about the body into the liver, the appendix, the stomach, and even up the oesphagus and out through the nose, to the horror of the surprised host.

Infection occurs chiefly among people with bad sanitary habits, though it may be obtained in the best establishments from eating inadequately washed fresh salad vegetables grown in soil contaminated with human faeces. The worms are common in the southeastern U.S.A. where the children, particularly, are likely to deposit their faeces in the yard adjoining the house. The eggs are spread around by pigs, by the family dog, or by the children themselves, who finally get the eggs on their hands and carry them into the kitchen.

An ascaris which cannot be distinguished by its structure from the one that lives in man is very common in pigs. However, the ascaris eggs from pigs do not ordinarily develop to maturity in man, nor do those from man's faeces infect the pig.

MUCH more serious than the big ascaris is the tiny HOOKWORM (*Necator americanus*, a name which means 'the American

killer'). The mouth cavity of the worm contains plates by which the worm grasps a
bit of the intestinal lining of the host and
holds on while it sucks in blood and tissue
fluids. The eggs pass out in the faeces and
fall on the ground, where they hatch into
larval worms. These live in the soil for
some time, feeding and growing. After
they have attained a certain size and have

HOOKWORMS, natural size.

stored up food, they cease to feed and are capable of infecting
man. They invade their human host by burrowing through his
skin, and infection most often occurs from the habit of going bare-
foot in localities where the soil is likely to contain human faeces.
After entering the skin, the worms pursue the same course as
described for the ascaris, eventually reaching the intestine.

Hookworm disease in the United States is confined largely to rural
parts of the southeastern states, where millions are affected. Adequate
moisture and temperatures between 68° and 86°F. are necessary for the
development of the worms in the soil; hence, hookworm disease of
man (and also of other mammals) is restricted to regions such as the
South, where these conditions prevail. The symptoms of the disease are
widely known: anaemia, laziness, and general lack of physical and
mental energy. These conditions lead to a retardation of physical and
mental development, so that an infected child of fifteen years of age
may appear to be only ten years old. The 'poor white trash' of the
South have suffered such inefficiency for generation after generation.
The resulting poverty, ignorance, and deterioration of culture only
accentuate the condition. The fact that these poor people have been held
in contempt by their more fortunate neighbours, who attribute the
condition to 'natural-born shiftlessness', has not been helpful. Negroes
harbour the worms but are not so susceptible to their harmful effects.
During the last two decades health agencies have done much to stop
this drain on the national economy and culture.

Simple treatment with drugs eliminates most of the worms, but in
addition to treatment it is necessary to prevent new worms from entering. Wearing shoes and avoiding contact of the skin with infected soil
is one factor. A second is the sanitary disposal of faeces which contain
eggs. Towards this end government agencies have constructed hundreds
of thousands of privies in the southern states.

THE TRICHINA WORM (*Trichinella spiralis*) is a much dreaded
parasite of man. It is usually obtained by eating insufficiently
cooked pork, but occasionally is contracted from other kinds of

meat – bear meat, for example. The worms become sexually mature in the human intestine. The eggs hatch in the uterus of the female, which thus gives birth directly to larval worms. The larvas gain access to the blood and lymph vessels of the intestine and are carried about through the body. They leave the vessels and burrow into the muscles – usually those of the diaphragm, ribs, tongue, and eyes. In the muscles the larvas increase about ten times in size, to a length of 1/25 inch, and then encyst, curling up and becoming enclosed in a thick wall formed by the host tissue. They develop no further and eventually die, unless the flesh (containing the cysts) is eaten by a suitable host. Pigs obtain the worms by eating the flesh of other animals, usually fragments of slaughtered pigs or rats, containing encysted larvas. In the body of the pig the worms go through the same history as recorded above for man. The muscles of an infected pig contain numerous encysted trichina worms; and if man eats infected pork that has not been thoroughly cooked a trichina infection usually results.

The adult worms do no harm, and after a few months disappear from the intestine. The greatest injury occurs during the migration of the larvas, when half a milliard of more of them may simultaneously bore through the body. At this time there are excruciating muscular pains, muscular disturbance and weakness, fever, anaemia, and swellings of various parts of the body. It is during this stage of the disease that death may occur, and about a third of the sufferers die. If the victim survives this period, the larvas become encysted and the symptoms subside, though there may be permanent damage to the muscles. In less heavily infected cases the symptoms may be mild and are likely to be diagnosed as 'intestinal trouble'. Serious cases are often diagnosed as typhoid fever. Thus, the actual occurrence of this disease is much higher than is generally supposed. Autopsies have shown that about 20 per cent of the population have suffered from trichinosis at some time or other.

It is probable that the religious laws of the Jews prohibiting the eating of pork resulted from experience with trichina infections, although at that time nothing was known about the worms themselves. At the present time the United States government does not inspect pork for the occurrence of encysted trichina worms, since such inspection requires microscopic examination, and light infections could be readily overlooked anyway. Inadequate inspection is worse than none, because it gives a false sense of security to the consumer. The absolute prevention of trichinosis lies with the consumer, who has only to cook all pork thoroughly. It is important not to roast pork in pieces so large that heat does not penetrate to the centre. Large public outdoor picnics are often a source of epidemic trichinosis. Many of the worst cases have been due to 'home-made' pork sausage which was improperly prepared. An ounce of infected pork sausage may contain one hundred thousand encysted worms. All market animals are parasitized in some way or other, and it is understood that the consumer will prepare his food properly to safeguard himself against infection.

MICROFILARIA in blood. The worm is encased in a transparent sheath, really the inner lining of the egg. Three red blood corpuscles are shown to give scale. (Modified after Faust.)

THE FILARIA (*Wuchereria bancrofti*) is a roundworm of great importance as a human parasite in tropical and subtropical countries. In the United States it occurs in only one locality, Charleston, South Carolina. This worm differs from the preceding ones in that an intermediate invertebrate host is involved in the life-cycle. The adult worms look like coiled strings as they lie in the lymph glands or ducts of an infected person. The female is 3 or 4 inches long, and the male about half this length. The female gives birth to small larvas, known as MICROFILARIAS, which get into the blood vessels and develop no further unless sucked up by a mosquito of the right species. Within the mosquito the larvas continue development and migrate to the biting apparatus. When the mosquito bites another person, the worms creep out onto the skin of the victim and penetrate near the bite. The chief consequence of filaria infection is the blocking of the lymph channels. This results in immense swelling and growth of the affected parts – a condition known as *elephantiasis*.

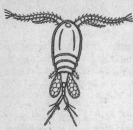

CYCLOPS, a one-eyed crustacean, serves as the intermediate host of the guinea worm.

THE GUINEA WORM (*Dracunculus medinensis*) is one of the more serious discomforts of life in India, Arabia, Egypt, and Central Africa. The male worm is not well known but is probably only about an inch long. The female is from 2 to 4 feet long and 1/25 inch in diameter. It usually lives under the skin but sometimes lies near the surface, where it appears like a coiled varicose vein. When mature, the female approaches the surface of the skin, usually that of the arms or legs, and secretes a toxic substance which causes a blister to form. The blister breaks, exposing a shallow depression or ulcer with a hole in its centre. When this ulcer is suddenly plunged into cold water (as by women when they wash clothes in the river), a milky fluid containing large numbers of tiny larvas is ejected from the hole in the ulcer. The larvas swim about in the water until they find a cyclops, a small crustacean, into which they enter and in which they undergo development. When man drinks unfiltered water containing a cyclops, the larvas are introduced into their final host.

In some places one-fourth of the population is incapacitated during part of the year by the guinea worm. Some of the symptoms, which appear at the time the blisters are formed, are vomiting, diarrhoea, and dizziness. Native medicine men usually extract the worm by slowly and painfully winding it out on a stick. This often results in infection, followed by loss of the limb or by death. This method is quite successful, however, if done by a doctor, who uses the proper precautions against infection. Control of the disease would be very easy if infected natives could be taught to stay out of the water and if communities could be induced to filter their drinking water. In India this is difficult because of the religious traditions that surround the ways in which the people obtain and use water.

HAIRWORMS

IT is an old belief that the so-called 'horsehair snakes' arise from horsehairs that have fallen in the water. The horsehair snakes are neither horsehairs nor snakes, but are members of the small phylum NEMATOMORPHA ('form of a thread'). It is not difficult to understand how the erroneous notion of their origin got its start when we consider that these worms, which live in almost any body of fresh water, are often found in drinking-troughs, and that many of them are about 6 inches in length, black or brown in colour, and, though somewhat thicker, look not unlike the hairs of a horse. Also, it seemed necessary to explain why one would see no trace of them on one day and then find numbers of these worms in the same place on the next day. We now know that this sudden appearance of the worms is due to the fact that the larvas develop as parasites in insects and the adults emerge full-grown from their insect hosts. They

probably drop into the water when their hosts approach a pond or stream, or perhaps are swept into the water by a heavy rain.

The hairworms resemble the roundworms in structure and sometimes are included as one of the classes of the phylum Nematoda. Their life-history differs from that of parasitic nematodes in that the adults are all free-living. The difference is not important, however, because the adults in many cases lack a mouth; and even those with a mouth probably do not feed. Thus, the free-living adult may be looked upon as only a reproductive stage, though it may sometimes last for months. The female lays the eggs in long egg-strings which she winds about water plants. The larvas that hatch have a spiny proboscis by means of which they bore their way into the body of an aquatic insect larva. The next stages are not well known, but we find the mature hairworms in the bodies of land beetles, crickets, and grasshoppers. Perhaps the transfer to the land host is effected when the aquatic larvas mature, go on land, and are eaten by a beetle; or the first insect host may die and its parasitic hairworm larvas escape and bore their way into the second host. In the body of the second insect host they develop into adults, which finally return to the water to mate and lay eggs.

Gordius (shown in the illustration at the top of this page) is a genus of hairworms which wriggle about in ponds

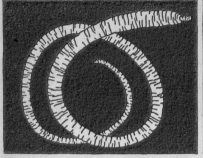

Spiny-headed worm from the intestine of a pig.

Head, enlarged. (Modified after Van Cleave.)

and ditches all over the world. The name comes from the fact that the adults are often found together in masses so tangled as to suggest a 'Gordian knot'.

SPINY-HEADED WORMS

THESE elongated, cylindrical worms live as parasites in the intestine of vertebrates and used to be considered as a class of nematodes. But as it is difficult to reconcile their unique body plan with that of round-worms or of any other group, they are now set aside by themselves as the phylum ACANTHOCEPHALA, a name that means 'spiny-headed' and refers to their most characteristic structure, an anterior retractile proboscis armed with rows of stout recurved hooks. Behind the proboscis is a short neck region and then the body proper, which is roughly cylindrical. By means of the burrlike proboscis the worm clings to the intestinal lining of its host, absorbing nourishment through the delicate cuticle. There is no trace of a digestive tract.

Acanthocephalids shed their eggs in the faeces of the host. If the host is an aquatic vertebrate, the eggs are probably eaten by a crustacean or an aquatic insect, and in these animals the larvas develop. They get back into a vertebrate when the intermediate host is eaten by the vertebrate final host. The life-history is similar for land vertebrates, but it involves land insects. A species common in rats and another which lives in pigs are both occasionally found in man. The rat parasite is from 2 to 10 inches long.

The eggs are shed in the faeces; and when rat faeces are eaten by cockroaches, the larvas develop. Rats become infected by eating cockroaches, which sometimes form their chief article of diet. Man probably becomes infected when he unwittingly eats an infected cockroach. The acanthocephalid of pigs is a huge worm over a foot long, with a pinkish wrinkled body. Pigs become infected by eating grubs (larvas of the June beetle) which they find as they root about in the soil.

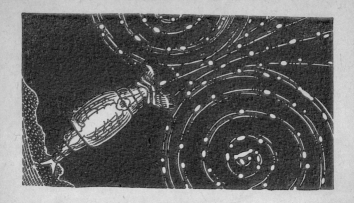

CHAPTER 16

Lesser Lights

THE animal kingdom is divided into about twenty phyla. The exact number depends upon how many different plans of organization the classifier thinks there are. Some of the phyla are more important than others – at least to man – and among those usually considered of less importance, six will be discussed in this chapter.

These owe their inclusion among the 'lesser lights' to one or all of the following reasons: they have a small number of species or of individuals; the members are of small size; they constitute no important source of food or of disease for man; and they illustrate no principle of theoretical interest that is not as well shown by other phyla.

ROTIFERS

WHENEVER a body of fresh water is examined for free-living protozoa, one is almost certain to find, in addition, microscopic animals about the size of protozoa but consisting of the equivalent of many very small cells and with a grade of structure a little more complicated than that of flatworms in some respects, less so in others. These are the rotifers. Because they are microscopic and play no important role in man's economy, these abundant animals are little known except to zoologists and amateur microscopists, who seldom fail to be fascinated by their great variety of shapes (many of them truly fantastic) and by their rapid and often seemingly incessant motions.

Rotifers can be recognized at once by the presence at the anterior end of a CROWN OF CILIA, which serves as the chief organ of locomotion and also as the means of bringing food to the mouth. In some forms the beating of the cilia, which are arranged around the edge of one or more disc-shaped lobes, gives the appearance of a revolving wheel – hence the name of the phylum, ROTIFERA, which means 'wheel-bearers'.

Rotifers vary in shape from wormlike bottom-dwellers, or

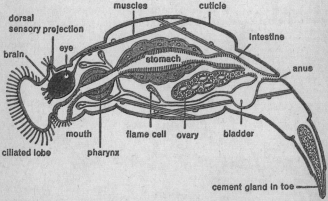

ROTIFER, showing structure. Only a few of the nuclei are shown (e.g., muscles and stomach wall). (Combined from several sources.)

flower-like attached types, to rotund forms that float near the surface; but all are bilaterally symmetrical. In many species (as in *Philodina,* shown in the illustration above the chapter heading) the body is elongated and is roughly distinguishable into three regions: a HEAD which bears the mouth and cilia; a main central portion, called the TRUNK; and a tapering portion, known as the FOOT. At the end of the foot are the 'toes', pointed projections from which open cement glands that secrete a sticky material used to anchor the rotifer during feeding. The toes aid in a second method of locomotion in which the animal proceeds 'inchworm fashion'. It stretches out, takes hold at the front end, releases the toes, and contracts the body; then it

The TEETH OF THE PHARYNX are the most distinctive structures of rotifers and are used as a basis upon which to distinguish one species from another. (After Harring and Myers.)

fastens the toes again, and extends, etc. The whole body is inclosed in a transparent, flexible cuticle, which is folded into sections that can be 'telescoped' one into the other when the animal contracts.

When FEEDING, a rotifer remains attached to a bit of debris, and the rapid beating of the cilia draws a current of water towards the mouth. Protozoans and microscopic algae are swept through the mouth into a muscular pharynx (or mastax), which contains a chewing apparatus consisting of little, hard 'teeth', operated by muscles. In some rotifers the long pincer-like teeth can be extended through the mouth and used, like a forceps, for catching prey. The pharynx leads into a straight digestive tract which opens by an anus at the junction of trunk and foot.

In GENERAL STRUCTURE a rotifer is similar to flatworms, nemerteans, and nematodes. The cuticle serves as a place of attachment for the muscles by which the animal moves. Muscular activity is co-ordinated by a simple nervous system which centres about a brain in the anterior end. Many rotifers have simple eyes and sensory projections called 'antennas'. The excretory system is like that of the flame-cell systems of flatworms and nemerteans; but in addition the terminal portion of the excretory tube is enlarged into a bladder which pulsates, ejecting its contents into the most posterior part of the intestine. There is no circulatory system in rotifers, and this is what one would expect of such minute organisms. Substances simply diffuse the microscopic distance from the gut to muscles and other tissues.

Rotifers are peculiar in that their bodies are not divided up into distinct cells but, like some of the tissues of

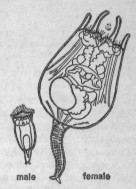

male   female

MALE ROTIFERS are usually small and degenerate. (After Hudson and Gosse.)

sponges, flatworms, and nematodes, consist of SYNCYTIA, that is, protoplasmic masses containing a number of nuclei. Cell walls are present in embryonic stages but later disappear. It is also a striking fact that the number of cells of the late embryo, or the number of nuclei of the adult, is constant (about one thousand) for each individual of a species; and, further, each nucleus occupies a definite position, so that all the nuclei of a rotifer can be numbered and mapped. Such cell constancy also appears to a limited extent in some other phyla.

The fertilized female-producing EGG OF A ROTIFER has a hard, thick shell which protects the egg during the resting period. (After H. Miller.)

Rotifers REPRODUCE sexually. The sexes are separate; but the males are generally small and degenerate, sometimes entirely lacking the digestive and excretory systems. Such individuals can live for only a few days. During most of the year, in the typical life-history, females give rise to other females by way of eggs that are not fertilized. This development of eggs without fertilization is called PARTHENOGENESIS and occurs also in other phyla. As the sexual season approaches, some of the females lay eggs which are smaller than and differ in other ways from the usual female-producing eggs. If not fertilized, these smaller eggs hatch into males. The males then fertilize the females, after which fertilized eggs are laid. These are distinguished from the parthenogenetic ones by a hard thick shell, often ornamented. They can withstand drying, freezing, and other unfavourable conditions, and after a resting period hatch into females. In one group of rotifers males have never been seen, and perhaps they do not occur. In this group the eggs develop without being fertilized and always become females.

Some rotifers can withstand DRYING even more than can many protozoans and nematodes. In this almost completely dried state they may live for years. As soon as moisture appears, they swim about and feed actively. Because of this capacity to resist drought, rotifers can live in places that are only temporarily wet, such as roof gutters, cemetery urns, rock crevices, among moss, and in similar places. When the water evaporates, the animal contracts to a minimum volume and loses most of its water content. Sometimes the animal itself dies, but its contained eggs survive until moisture returns. There are some marine rotifers, but the group is much more abundant in fresh water. Because of their small size and their capacity for withstanding temporary drying, rotifers have been distributed the world over, chiefly by wind and birds. If environmental conditions are similar, a lake in Africa will contain the same species of rotifers as a lake in North America.

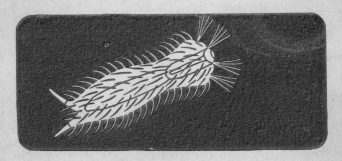

GASTROTRICHS

ALMOST any aquatic debris that contains rotifers will also contain a few members of the small phylum GASTROTRICHA. These minute many-celled animals are about the size of rotifers and resemble them in many details of structure. They have no crown of cilia but swim by means of tracts of cilia on the ventral surface. The cuticle is often clothed with scales or bristles, and gastrotrichs are likely to be confused with ciliated protozoans.

The digestive system is a straight tube with a muscular sucking pharynx more like that of nematodes than of rotifers. In the species commonly seen in fresh water the tail end of the body is forked; and at the tip of each fork is the opening of a cement gland, which serves the same function as in rotifers.

About one-quarter of the known gastrotrichs live in the ocean; these are hermaphroditic. The rest live in fresh water, and these are females which apparently reproduce parthenogenetically. No males have ever been seen.

BRYOZOANS

SOME of the small and more delicate 'seaweeds' admired by visitors to the seacoast are not seaweeds at all but the branching colonies of members of the phylum BRYOZOA, a name that

means 'moss animals' and refers to the plantlike appearance of many bryozoans. Because of the colonial habit of its members, some prefer to call the phylum POLYZOA ('many animals'). Some colonies are shrublike and hang from blades of kelp or grow out from crevices of rocks; others form flat incrusting growths on seaweeds and rocks; and some fresh-water bryozoans grow as gelatinous masses around stems and twigs that have fallen into the water.

Each individual of a colony lives in a protective case of hard material, calcareous or horny, into which it can withdraw completely. (The fresh-water bryozoan colony shown in the heading is *Plumatella,* which has a delicate and transparent horny covering.) At first glance the animals resemble hydroids, for at the anterior end they have a set of tentacles borne on a circular or horseshoe-shaped ridge, called the 'lophophore'. However, they are considerably advanced over the hydroids and have a grade of structure more like that of rotifers. Although, as in hydroids, the members of a bryozoan colony arise from one another by budding, they conduct their activities independently. When undisturbed, the animals emerge 'cautiously' and spread their tentacles in the water; but at the slightest vibration they retreat into their cases.

The tentacles are ciliated and, when spread in the water, create currents which drive microscopic organisms into a mouth situated within the ring of tentacles. The food is moved through the U-shaped digestive tract by means of cilia. The anus opens near the mouth, but just outside the circle of tentacles. The proximity of mouth and anus does not seem to us particularly desirable, but apparently it is a satisfactory adjustment for an animal that lives in a case with only one main opening.

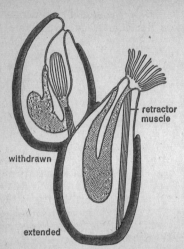

retractor
muscle

withdrawn

extended

TWO MEMBERS OF A BRYOZOAN COLONY, one
extended and one withdrawn. (Modified after
Delage and Herouard.

The hard case that enclo-
ses a bryozoan is secreted by
the underlying ectoderm.
This ectoderm and a layer
of mesodermal cells constitute the thin body wall. Between the body wall and the
digestive tract there is a
large fluid-filled space, completely lined with mesoderm.
Such a body cavity lined
with mesoderm is called a
COELOM ('hollow'), and the
coelomic lining is called the
PERITONEUM.

The bryozoans show a re-
markable reorganization at
various times, when the tentacles, gut, and other internal
organs degenerate, forming
a compact mass known as
the 'brown body'. New organs are regenerated from the body
wall; and the brown body, which comes to lie in the stomach of
the regenerated individual, is eliminated through the anus. Since
these bryozoans lack an excretory system, it is possible that the
formation of the brown body is related to excretion.

Some bryozoans illustrate POLYMORPHISM. In these we find,
attached to the normal feeding individuals, highly specialized
individuals which resemble a bird's head and so are called
'avicularia'. Each avicularium has a pair of jaws, operated by
muscles, which can snap shut upon any small animal that
wanders over, or chances to alight on, the colony. Presumably,
the function of these individuals is not to aid in feeding but to
prevent larvas (and other small animals) from settling upon, and
interfering with, the feeding activities of the colony.

Growth of the colony is by asexual budding. New colonies
are provided for by sexual reproduction. The ovaries arise from
the peritoneum of the body wall. The testes usually form on the
peritoneum covering the strand of mesenchyme-like cells that
fastens the intestine to the body wall. Eggs and sperms are shed
into the coelom, where fertilization takes place. The ciliated

larva of marine bryozoans is free-swimming and resembles a TROCHOPHORE, a kind of larva found in many invertebrates (see chaps. 17 and 19).

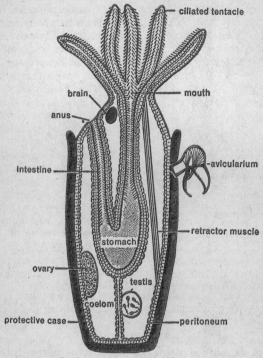

BRYOZOAN, showing structure. (Combined from several sources.)

Fresh-water animals usually do not have free-swimming larvas. Freshwater bryozoans have, instead, buds known as STATO-BLASTS. These, like the gemmules of fresh-water sponges, consist of a mass of cells surrounded by a protective covering. They survive the winter and develop into new bryozoan individuals.

THERE are two groups of bryozoans, which differ in a number of important ways. Although it is customary to group them in one phylum, they might properly be assigned to separate phyla. The description above applies to the more advanced group known as the ECTOPROCTA (meaning 'outside anus'), in which, as we have seen, the anus opens outside the circle of tentacles. In the ENDOPROCTA the anus opens within the ring of tentacles. Unlike the ectoprocts, which have no special excretory system, the endoprocts have a flame-cell system. In the endoprocts mesenchyme fills the space between the gut and the body wall, and there is no coelom. This has been taken as an indication that the

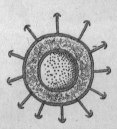

BRYOZOAN LARVA. (Modified after Prouho.)

The STATOBLAST of *Pectinarella*, a common freshwater bryozoan, is about $\frac{1}{32}$ inch in diameter and has a row of anchor-shaped hooks. (After Kraepalin.)

endoprocts are more primitive than the ectoprocts, for a coelom, unless secondarily reduced, is present in all higher phyla.

BRACHIOPODS

ONE of the early investigators, who pried open the shells of a brachiopod and looked inside, thought that the two spirally coiled ridges within the shell were 'arms' by which the animal moved and that they corresponded to the foot of a clam. From this mistaken notion came the name of the phylum, BRACHIOPODA, which means 'arm-footed'. The shells of a clam are right and left, while those of a brachiopod represent dorsal and ventral surfaces. The gape of the brachiopod shell is at the anterior end and the hinge is at the posterior end; the shells can be opened and closed by means of muscles. The posterior end of the body is extended into a stout muscular stalk (peduncle) by which the

animal is attached, usually to a rock. The shells are secreted by two folds of skin which enclose the main part of the body.

Within the shells the most conspicuous structures are the two spirally coiled tentacular arms, or LOPHOPHORE, which are thought to correspond to the lophophore of a bryozoan. Running along each arm is a ciliated groove, and on one side of this a row of ciliated tentacles. The beat of the cilia sweeps minute organisms into a mouth situated between the bases of the arms, from there into a stomach supplied with digestive glands, and finally into an intestine. The water currents also maintain a steady supply of oxygen.

The circulatory system is simple and contains a contractile

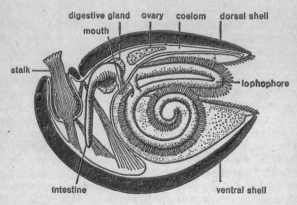

BRACHIOPOD, showing structure. (Modified after Delage and Herouard.)

muscular enlargement called the 'heart'. As in ectoproct bryozoans, there is a true coelom lined with mesoderm. The excretory tubes lead from the coelom to the large cavity in which the arms lie and which connects directly with the outside.

The sexes are usually separate, and the sex organs lie near the intestine. The eggs or sperms are discharged into the coelom and reach the outside through the excretory tubes. The ciliated larva resembles a trochophore.

There are two main groups of brachiopods. In the more primitive group, of which *Lingula* is an example, the two shells are of a horny texture, somewhat rectangular in shape, and of equal size. They are held together only by muscles; there is no hinge. The stalk is usually very long and passes out between the shells. In the more advanced and larger group of brachiopods, represented both in the illustration and in the diagram, on the preceding page, the shells are calcareous and are hinged together. The ventral one is larger than the dorsal, and at its posterior end has a kind of upturned 'beak' through which the short stalk passes. The dorsal shell bears two calcareous coiled projections which serve as an internal support for the tentacular arms. In this group there is no anus.

Brachiopods are all marine and are not very widespread. Only about two hundred living species are known. But there was a time in past geological ages when there were at least three thousand species, of which we now have good fossil records. At that time brachiopods played a very important role in the invertebrate world, comparable to that of the clams and oysters of the present.

Modern species of *Lingula* are almost identical with species which we estimate, from the fossil record, to have lived almost 500,000,000 years ago. This is a record for conservatism among animals, and *Lingula* has the 'honour' of being the oldest known animal genus. Fossil brachiopods are of great value to

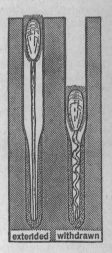

LINGULA lives in vertical burrows in the sand, attached to the bottom by the long stalk. (Modified after Francois.)

geologists, as they constitute one of the most important criteria for dating rock strata. (See also chap. 27.)

PHORONIDEA

THE phylum PHORONIDEA is a small one, consisting of only about a dozen species. These wormlike animals are all marine, sedentary, and tube-dwelling. A common species of the American West Coast lives in straight cylindrical tubes embedded vertically just below the surface of the substratum in mud and sand flats. The animals have a horseshoe-shaped lophophore, spirally coiled at the ends, which bears ciliated tentacles that catch food. In this food-catching organ, in the U-shaped intestine, and in other respects phoronideans resemble bryozoans. The larva is of the trochophore type.

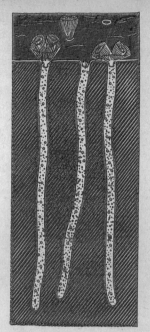

ARROW WORMS

IN the open ocean we find transparent, slender animals, usually 1–3 inches long, that look like cellophane arrows as they dart after their prey. Though at certain seasons they occur in incredible numbers, and at such times form a large part of the food of fish, the arrow worms are members of a phylum, the CHAETOGNATHA, which has relatively few species. The name means 'bristle-jawed' and refers to the curved bristles, on either side of the mouth, that aid in catching prey. The body is divided into head, trunk, and tail and has finlike projections, which probably serve as balancers. The brain is well developed, and there is a set of eyes. The anus is situated at the junction of trunk and tail, about a third of the way from the posterior end. The three body regions are separated internally by transverse partitions, and there is also a longitudinal partition which separates the coelom into

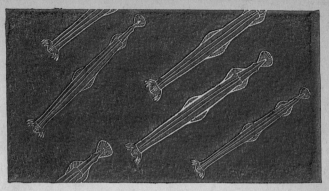

right and left halves. The animals are hermaphroditic: both male and female sex cells arise from the lining of the coelom. The body plan is so different from that of the other groups that it is difficult to say what relationships they have to other invertebrates. In certain details of development the chaetognaths resemble some of the members of the phylum to which man belongs.

CHAPTER 17

Soft-bodied Animals

THE second largest and second most familiar invertebrate group is the phylum MOLLUSCA. The name means 'soft-bodied'; and because of their soft, fleshy bodies, which are of relatively large size, the molluscs, more than any other invertebrates, are widely used as food by man. Some of the better-known molluscs are snails and slugs, clams and oysters, octopuses and squids.

Despite the lack of similarity in the external appearance of a snail, a clam, and a squid, their body plan is fundamentally the same and differs radically from those of all of the other inverte-brate groups. The typical features of a mollusc are much modi-fied, and some are even lost, in a highly specialized animal like a clam. They are less changed – from what we think was the condi-tion of the primitive molluscan ancestor – in the chitons.

THE CHITONS are sluggish animals which browse on the algae that grow on rocks near the seashore. When disturbed, they clamp down upon the rock so tenaciously with their powerful muscles that it takes much persistence – and often a chisel – to pry them loose.

The body is bilaterally symmetrical. At the anterior end is a reduced and inconspicuous head, which is probably a secondary adaptation to a sedentary life; it is the chiton's main disqualifying character as a typical mollusc. The primitive molluscan ancestor of the chiton probably had a prominent head with sense organs – more like a snail's head than that of a chiton. The ventral surface is largely taken up by a broad, flat muscular creeping FOOT, abundantly supplied with a slimy secretion. The VISCERAL MASS (containing most of the organs) lies dorsal to the foot and is completely covered by a heavy fold of tissue which extends around on each side of the foot, much as a roof covers a barn. This fold is called the MANTLE, and the part under the 'eaves' is the mantle cavity. On its upper surface, the mantle secretes a SHELL, which in most of the chitons consists of eight

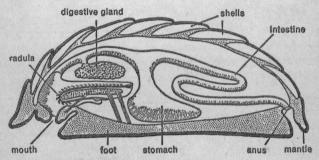

CHITON, showing principal structures of a relatively primitive mollusc. The gills are not shown (see the photographs at the end of chapter 18).

separate plates, overlapping from front to rear like the shingles on a roof. Between the mantle and the foot in the mantle cavity on both sides is a row of GILLS, thin-walled structures used in breathing.

The DIGESTIVE SYSTEM is a tube extending from the mouth, in the head, to the anus, at the posterior end of the animal. The mouth leads into a muscular chamber, the pharynx, in which is found the RADULA. This is a horny ribbon covered with many rows of hard recurved teeth. A complicated array of muscles pulls the radula back and forth over a cartilaginous projection, much as a cloth is pulled over a shoe in polishing it. When feeding, the chiton protrudes the radular apparatus through the mouth; and as the teeth of the radula move over the surface of plants, they rasp off small fragments. Behind the radula the oesophagus opens into the stomach, from which a long intestine runs to the anus, at the posterior end.

The CIRCULATORY SYSTEM is better developed than that of nemerteans. There is a specialized pumping organ, the HEART, and extensively branched blood vessels which carry blood from the heart to all parts of the body and then back again. The heart lies in a cavity, the PERICARDIAL CAVITY, which is a part of the body cavity, or coelom.

There are two EXCRETORY ORGANS (the kidneys) consisting of a glandular epithelium which extracts nitrogenous wastes from the blood passing through them. The waste material is discharged to the outside by way of pores near the anus.

The NERVOUS SYSTEM is a ring of nervous tissue around the oesophagus, connected with two pairs of longitudinal nerve cords which go to the muscles of the foot and mantle. It is a 'ladder type' of nervous system, not very different from that of nemerteans.

Eggs and sperms are shed into the sea water. The fertilized egg develops into a TROCHOPHORE LARVA, similar to that of several phyla already mentioned and also to that of the marine annelids (discussed in chapters 19 and 20). The typical trochophore is a spherical larva with a prominent band of cilia about the equator that serves as a locomotor organ and also to bring food to the mouth. Internally, the trochophore has a stomach and an intestine which connects with the exterior through an anus. There also develops a larval excretory organ with flame cells; it disappears when the larva changes into the adult. At the top pole of the larva is a group of sensory cells connected to a tuft of long cilia.

THOUGH less common than their more conspicuous and

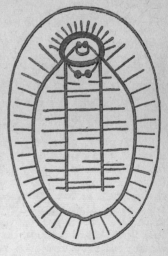

NERVOUS SYSTEM of the chiton.
(Based on several sources.)

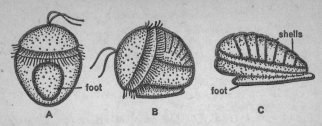

DEVELOPMENT OF A CHITON. A, trochophore with foot beginning to appear. B, larva attached by foot; shells starting to develop. C, shells clearly marked; adult appearance indicated. (After Heath.)

more economically valuable relatives, the chitons are described here because they display the molluscan body plan in its most typical form. The body consists of three main regions: a ventral muscular foot, a dorsal mass containing the viscera, and covering this a fleshy mantle which secretes the protective shell. A unique molluscan character is the radula. Not all molluscs have a radula, but nothing like it is found anywhere else in the animal kingdom. A shell is not peculiar to molluscs, for hard protective coverings have been developed by many groups. In most of these the outer covering seriously limits the activities of the animal. In molluscs there is a compromise, and the shell does not always completely encase the animal. In a snail, for instance, the head and foot can be extended when the animal is moving, withdrawn when danger threatens. The success of the molluscan plan is attested by the fact that there are over seventy thousand species of molluscs – ten thousand more than in the phylum to which man belongs. How the various kinds of molluscs have adapted the same body plan to their specialized and very different ways of life is the subject of the next chapter.

MORE ABOUT PENGUINS
AND PELICANS

Penguinews, which appears every month, contains details of all the new books issued by Penguins as they are published. From time to time it is supplemented by *Penguins in Print*, which is a complete list of all books published by Penguins which are in print. (There are well over three thousand of these.)

A specimen copy of *Penguinews* will be sent to you free on request, and you can become a subscriber for the price of the postage. For a year's issues (including the complete lists) please send 30p if you live in the United Kingdom, or 60p if you live elsewhere. Just write to Dept EP, Penguin Books Ltd, Harmondsworth, Middlesex, enclosing a cheque or postal order, and your name will be added to the mailing list.

Some other books published by Penguins are described on the following pages.

Note: *Penguinews* and *Penguins in Print* are not available in the U.S.A. or Canada

THE THEORY OF EVOLUTION

John Maynard-Smith

All living plants and animals, including man, are the modified descendants of one or a few simple living things. A hundred years ago Darwin and Wallace, in their theory of natural selection, or the survival of the fittest, explained how evolution could have happened, in terms of processes known to take place today. This revised edition describes how their theory has been confirmed, but at the same time transformed, by recent research, and in particular by the discovery of the laws of inheritance.

After stating the problem and Darwin's answer to it, the author describes what can be learnt from laboratory experiments, and then gives the evidence that evolutionary changes are taking place today in wild populations. Later chapters discuss the origins of species, and the special problems which arise in studying the origins of major groups of animals and plants. The book ends by contrasting evolutionary and historical changes, and considers the relative importance of the two processes in the origin and future development of human society.

INTRODUCING BIOLOGY

James F. Riley

Biology, the science of life, embraces the growth, structure, and functioning of organisms, animals, and plants, the how and why of our bodies, and man's position in the universe of nature. And recent advances in biology have been no less dramatic than in other branches of science. Aided by the electron microscope the experimental biologist, long content just to analyse the living processes, is now poised for the artificial synthesis of life.

In this brief introductory outline of our knowledge of man and his environment, Dr Riley's themes are the workings of evolution and natural selection. He employs the lives and characters of the great biologists to disclose the progress of discovery in this field and build up a stimulating (and painless) introduction to biology, which will appeal to career and leisure students alike.